**Monographs in Electrical and Electronic
Engineering · 28**

Series Editors: P. Hammond, T. J. E. Miller, and S. Yamamura

Monographs in Electrical and Electronic Engineering

An Introduction to Ultrasonic Motors

Toshiiku Sashida

Shinsei Industries
Tokyo, Japan

and

Takashi Kenjo

Department of Electrical Engineering and Power Electronics
University of Industrial Technology
Kanagawa, Japan

CLARENDON PRESS • OXFORD
1993

Oxford University Press, Walton Street, Oxford OX2 6DP
Oxford New York Toronto
Delhi Bombay Calcutta Madras Karachi
Kuala Lumpur Singapore Hong Kong Tokyo
Nairobi Dar es Salaam Cape Town
Melbourne Auckland Madrid
and associated companies in
Berlin Ibadan

Oxford is a trade mark of Oxford University Press

Published in the United States
by Oxford University Press Inc., New York

Published in Japan
by Sogo Electronics Publishing Co.,
2-5-2 Sarugaku-cho, Chiyoda-ku, Tokyo

A catalogue record for this book is available from the British Library

Library of Congress Cataloging in Publication Data
Sashida, Toshiiku.
An introduction to ultrasonic motors / Toshiiku Sashida and
Takashi Kenjo.
(Monographs in electrical electronic engineering: 28)
Translated from Japanese.
Includes bibliographical references and index.
1. Ultrasonic motors. 2. Piezoelectricity. I. Kenjo, Takashi.
II. Title. III. Series.
TK2537.S27 1993   621.46–dc20   93–7171
ISBN 0-19-856395-7

Typeset by
Colset Pte Ltd., Singapore

Printed in Great Britain by
Biddles Ltd,
Guildford & King's Lynn

Preface

My first visit to Toshiiku Sashida's laboratory on the outskirts of Tokyo took place on a hot summer afternoon in 1981. After demonstrating a fan powered by his newly developed motor, he explained the motor's principles to me in clear, concise terms. When he had finished explaining, I had the distinct impression that I had met an extraordinary physicist. By the time I made my second visit the following year, he had already completed his next invention, the travelling-wave motor. Once again, as I sat listening to him explain the conceptual development which led to the new motor, I saw a streak of genius in the intuitive grasp he had for the physics of wave phenomena.

As a graduate student, I had done research on a wave phenomenon which takes place in ionized gas, called moving striations. I also recall lectures on flexural waves and on the piezoelectric effect in quartz. Yet it had never occurred to me to make the necessary link between elastic waves and piezoelectric phenomena to formulate a concept for the ultrasonic motor. It took an innovative mind like Sashida's to do so.

In February 1983, Sashida gave a special lecture at Motortech Japan which was sponsored by JMA (Japan Management Association) and chaired by myself. The auditorium was filled with an over-capacity crowd who stayed rapt throughout the lecture and demonstration. This lecture triggered a multitude of ultrasonic motor research programmes in corporate laboratories and universities throughout Japan. Among these institutions, Canon is probably the best known, with the motor's application for the autofocusing mechanism of its camera.

Yet it took close to ten years before the motor's applications started appearing in various fields. During this period, physicist Sashida continued his own work to develop a commercial prototype motor, while running a modest lapping-machine factory. When in 1990 he announced his intention to withdraw from lapping machines so that he could concentrate on ultrasonic motors, I also made a personal decision. This was to help organize the motor's principles and design theory into a unified body of knowledge so that it could be published as a monograph to be read by scientists and engineers around the world.

Although co-authorship was not in my mind at the beginning, I soon realized that a personal commitment was necessary to carry through such a project, and in this sense, having had prior publishing experience with OUP was an important deciding factor. Since I was working on another manuscript at the time, I waited until its completion before proposing the joint writing project to Sashida (to which he agreed immediately). We then decided that the book should first be written in Japanese, then translated later. I knew from prior experience that to attempt to write a book in English, a second language to us, from the outset would have been inefficient and delayed its publication.

The first Japanese edition was published in February 1991. Since I had other publishing commitments for several works in Japan, the translating work and revising necessary for the present version inevitably got delayed. We thus decided to hire a professional translator so that we could concentrate on updating and revising the contents. Fortunately, we were put in touch with a competent translator, Mr R. Takeguchi, who besides using his translating skills, followed through the nearly 300 equations to make sure he correctly understood the subject matter.

I was also peripherally involved in negotiations to publish Ueha and Tomikawa's *Ultrasonic motors: theory and applications* from OUP, which should appear shortly after this work. Whereas the present volume mainly discusses Sashida's invention and some of my theories concerning electrical machinery, the aforementioned work discusses ultrasonic motor developments in Japan subsequent to Sashida's 1983 lecture. Thus the two books complement each other in content.

Both Sashida and I wish to express our gratitude to our editor, Mr Richard Lawrence, who was encouraging from the very beginning, and to Mr T. Takano whose tireless efforts made the publication of the Japanese edition possible. Finally, I wish to thank Mr T. Kato, a former student of mine, who did the illustrations for this book.

Tokyo
February 1992   T. K.

Contents

1. What is an ultrasonic motor?

We shall first define the ultrasonic motor, then discuss some of the earlier researches which preceded the invention by Sashida. An ultrasonic motor is a type of actuator that uses mechanical vibrations in the ultrasonic range as its drive source. Human ears are capable of detecting sound waves from 50 Hz to 20 kHz. This is called the audible frequency range. Ultrasonic waves are sound waves or mechanical vibrations with frequencies above 20 kHz.

The use of ultrasonic waves is a direct outcome of using piezoelectric ceramics, which expand or contract in accordance with an applied electric field, as the vibrational source. In the ultrasonic motor, a voltage is applied to the piezoelectric ceramic element to generate alternating expansions and contractions, either in the ceramic body itself or in an attached metal piece. This will be discussed in Chapter 3. The magnitude of these oscillations is extremely small, of the order of 1 μm. In order to obtain a higher gain, the resonance effect of the ceramics in the ultrasonic range is utilized.

The oscillations are mechanically rectified in the motor to obtain unidirectional movement. Although the magnitude of an individual cycle of the movement is of the order of micrometres, it is possible to obtain high speeds, owing to the high frequencies (tens of kilohertz).

1.1 Inventions prior to Sashida

Attempts to obtain power from the considerable energy of ultrasonic vibrations generated by magnetostriction or electrostriction (or piezoelectric) vibrators began relatively early. It was known that the theoretical energy density of vibrators using piezoelectric ceramics was as high as several hundred watts per square centimetre, which is five to ten times that of electromagnetic motors.

The earliest research into actuators using vibrational energy as their drive source was conducted by H. V. Barth[1] with his ultrasonically driven motor. Shown in Fig. 1.1, this motor consists of a rotor and two vibrating drive units. The rotor moves clockwise when vibrator 1 is excited, counterclockwise when vibrator 2 is excited.

V. A. Gromakovskii and his colleagues[2] proposed a motor in which the rotor's movement is caused by exciting a piezoelectric vibrator, as shown in Fig. 1.2. When the vibrator is excited in the longitudinal direction at high speed, it collides with the rotor's surface as it expands and travels a distance δ. At the same time, the tip is displaced in the transverse direction

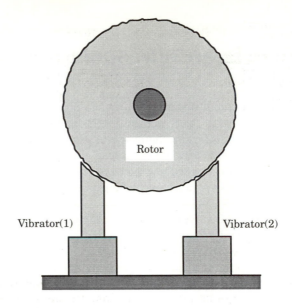

FIG. 1.1.   Principle of H. V. Barth's ultrasonically driven motor.

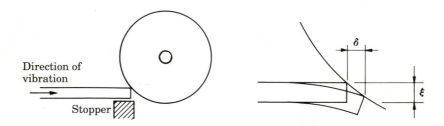

FIG. 1.2.   Principle of the piezoeletric motor of V. A. Gromakovskii *et al.*

by a distance ξ causing the rotor to move a distance $\sqrt{(\delta^2 + \xi^2)}$ at its surface.

1.2   Sashida's wedge-type ultrasonic motor

Sashida developed his first ultrasonic motor by using the two sources cited above as references and improving them. Its construction is shown in Fig. 1.3. We shall call this the wedge-type motor. The vibrator used is the bolt-tightened Langevin vibrator, which is named after French physicist Paul Langevin and will be discussed in detail in Section 3.5. This motor

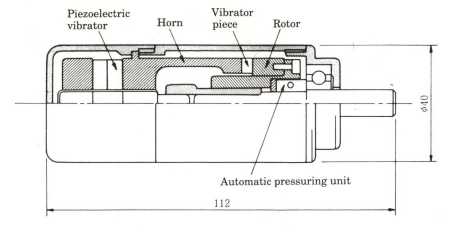

Piezoelectric vibrator   Horn   Vibrator piece   Rotor

$\phi 40$

Automatic pressuring unit

112

FIG. 1.3. Cross-section of the wedge-type motor.

is a significant improvement over previous models and is illustrated in Fig. 1.4. The vibrator's end moves in the longitudinal direction, to which a wedge-shaped vibrator piece P is attached. The tip of the vibrator piece collides with the rotor disc's surface, which is slightly angled (6°) from the vertical. The vibrator-piece tip is then forced to move vertically when it collides, and oscillates in resonance with the rotor's movement. The tip thus describes an elliptical path. When the tip moves up, it engages the rotor through friction and transmits motion. When moving down, there is no contact with the rotor's surface and no motion is transmitted. During the latter phase, the rotor will continue its rotation in the upward direction, owing to its inertia.

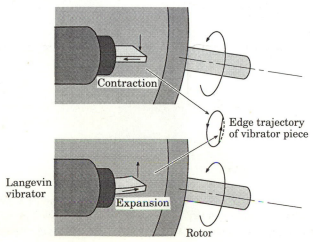

Contraction

Edge trajectory of vibrator piece

Langevin vibrator

Expansion

Rotor

FIG. 1.4. A wedge effect is used to prevent sliding in Sashida's method.

In the rudimentary motors shown in Figs 1.1 and 1.2, the longitudinal movement of the vibrator was applied directly, causing a large amount of sliding at the contact surface. Sliding causes great frictional heat loss, which made these models impractical.

The main feature of Sashida's method is the use of resonance and contacting at an angle near to a right angle, which considerably reduces sliding. The tip of the vibrator piece is cut at a slight angle (or wedge-shaped) so that firm contact is made with the rotor to prevent sliding. The tangential reaction at the rotor surface, whenever a collision occurs, creates resonance in the vibrator tip, and repeated contacts create a continuous rotary motion.

A photograph of a fan being driven by a prototype motor is shown in Fig. 1.5. The motor's characteristics are shown in Fig. 1.6. The torque-speed relationship is similar to that of an a.c. eddy-current motor. From this graph, it can be seen that the wedge-type motor has several features, which are described below.

(1) *High speed*. The motor has a no-load speed of 3000 rev min^{-1}, which is created by a two-step amplification process: the piezoelectric ceramic's vibration is first amplified several hundred times by the Langevin vibrator, then the longitudinal movement is converted to a tangential one, further contributing to increase the speed.

(2) *High efficiency*. The maximum efficiency of this prototype model is 60 per cent, which is higher than for an a.c. motor of the same size. The results of basic experiments with the wedge-type motor are discussed in

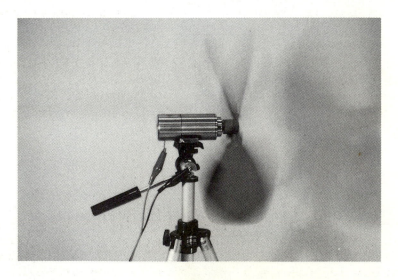

Fig. 1.5.   A fan powered by the wedge-type motor.

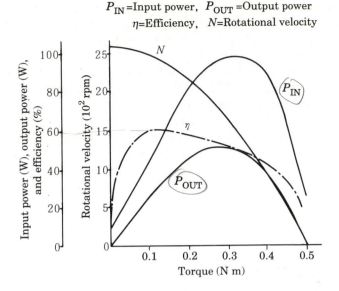

P_{IN}=Input power, P_{OUT} =Output power
η=Efficiency, N=Rotational velocity

FIG. 1.6. Characteristic of the wedge-type ultrasonic motor (example).

Chapter 4. In these experiments, a maximum efficiency of 87 per cent was obtained. This was possible through the use of only one vibrator piece and the careful attention given to such items as fine adjustment of the resonant frequency and the precision of the parts. The wedge-shaped vibrator configuration achieved high efficiencies which were not theoretically possible in previous models. The motor in Fig. 1.3, however, uses multiple vibrator pieces and so does not have such a high efficiency.

(3) *Short life span*. The wedge-type motor's biggest drawback is the rapid wear caused by friction at the contact surface between the vibrator and rotor. Although slippage is minimized by positioning these parts so that contact is made at almost a right angle, a small amount still occurs immediately after contact and just before separation, resulting in wear. For this reason, the motor has a relatively short life span.

1.3   The travelling-wave ultrasonic motor

It was when Sashida switched his thinking from the wedge-type to the travelling-wave motor around 1982 that the ultrasonic motor began attracting considerable attention in Japan. Since then, many Japanese firms have actively taken up research in this area and numerous patent applications have been filed. Following industry's lead a few universities are now also conducting research.

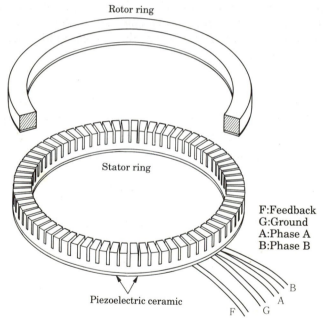

Rotor ring

Stator ring

F:Feedback
G:Ground
A:Phase A
B:Phase B

Piezoelectric ceramic

B
A
F   G

Fig. 1.7. Interior view and major components of the travelling-wave ultrasonic motor.

Figure 1.7 shows a photograph of the interior of a ring-type travelling-wave ultrasonic motor and an illustration of its major components. To see how the ultrasonic motor works, we need a basic knowledge of wave phenomena. The following section is devoted to this.

1.3.1 Standing waves and travelling waves

Consider a stretched string as shown in Fig. 1.8. When plucked, it will vibrate, with its outline following the arc of a bow. Every point on the string vibrates sideways. Although the vibrational frequency is the same at all points, the amplitude varies with position. This type of wave is known as a standing wave.

Another type of wave is the travelling wave. Most of us have the experience of throwing a stone into a pond and seeing ripples of waves spread outward in concentric circles. The ripples are travelling waves, that is, waves which propagate. A travelling wave transmits energy. As the stone hits the water's surface, part of its kinetic energy is converted into wave energy. This wave energy has a tendency to spread out and dissipate. The velocity at which the wave advances is called the phase velocity. In general, the velocity of energy propagation differs from the phase velocity. It is the wave energy which acts when, for example, a floating leaf is thrown out of the wave when the ripples reach the edge of the pond. Such waves are utilized in a travelling-wave motor.

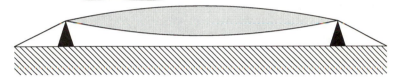

FIG. 1.8. Standing wave in a stretched string.

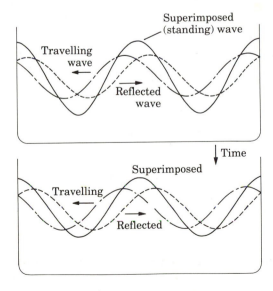

FIG. 1.9. The standing wave which appears on the water surface is created by superimposing the travelling wave and the reflected wave.

Travelling waves and standing waves are closely related. Consider, for example, a large tub filled with water. If one applies a bowl to the centre and then quickly lifts it away, standing waves are created. As shown in Fig. 1.9, a travelling wave hits the side of the tub, then is reflected and travels in the opposite direction. The standing wave can be viewed as the superimposed result of travelling waves going in opposite directions. In a stretched string, standing waves are created by waves reflected at the string ends.

1.3.2   Rayleigh wave and flexural wave

When part of a metal bar is vibrated, travelling waves are generated and propagate in both directions. Such waves can be classified into two general types. One, the Rayleigh wave, creates large amplitudes at the surface (as in earthquakes). Lord Rayleigh was the first to identify the characteristics of such waves while conducting research on earthquakes (see Fig. 1.10). Large amplitudes are created at the surface of the medium, which diminish exponentially with depth. Rayleigh waves in steel and bronze, when they are excited at a frequency of 40 kHz, have wavelengths of 7.5 and 4.75 cm respectively. Such wavelengths are too large for use in motors.

The other type of wave, known as the flexural wave, propagates with a snake-like motion (Fig. 1.11). For both Rayleigh and flexural waves, a point on the surface will move in an elliptical trajectory, instead of a simple up-and-down pattern. It is this elliptical motion that provides the drive in the travelling-wave motor.

1.3.3   A comparison with waves in water

Consider waves surging on to a beach. If a piece of wood floating among the waves is observed closely, it will be noticed that it moves in an ellipse. In sea waves, the crest moves in towards the beach, while the trough moves

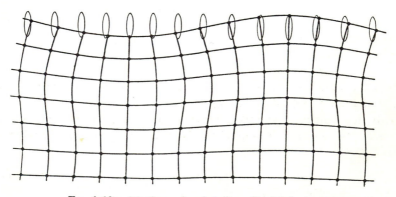

Fᴵɢ. **1.10.**   Motion of points in a Rayleigh wave.

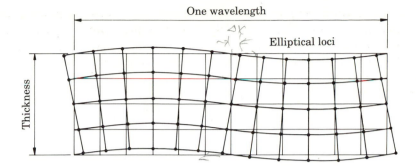

One wavelength

Elliptical loci

Thickness

FIG. **1.11.**   Motion of points in a beam.

out towards the sea. That is, the direction in which the wave travels and that of the wave crest's motion coincide. In metal bars, however, they oppose one another. Thus we see that waves in water and elastic waves in metal, although similar, are not quite the same. We shall leave a detailed and mathematical treatment of flexural waves for Chapter 5.

In the ultrasonic motor, travelling waves are used instead of standing waves. Moreover, the waves must be unidirectional. If a straight bar of finite length is used, the waves are reflected at the fixed ends and standing waves are created. So to achieve the same effect as an infinite bar to create travelling waves, the ends are connected to form a ring. By mounting two or three sources of vibrations at appropriate intervals we can create waves travelling in only one direction.

Although Fig. 1.12 illustrates the principles of a linear motor (i.e. moving in a straight line), it can also be viewed as a section of a ring in a rotary motor. Here, two piezoelectric ceramic elements are used as vibrators and positioned at an appropriate distance from one another. One of them, A, will generate a vibration given by

$$C \sin \omega t \tag{1.1}$$

while B, the other element, will generate a vibration of

$$C \cos \omega t \tag{1.2}$$

where C is the amplitude of the vibration and ω is the angular frequency (i.e. frequency f multiplied by 2π).

With this set-up, a travelling wave propagating to the right will be generated (shown near the top of the diagram). If B generates a vibration given by

$$-C \cos \omega t \tag{1.3}$$

the direction of the travelling wave will be reversed.

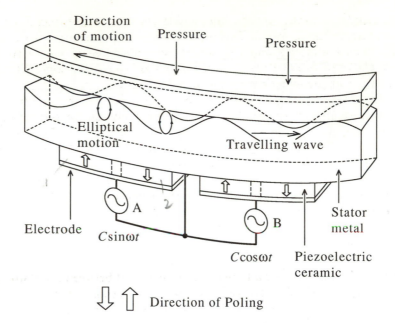

FIG. 1.12.   Principle of the travelling-wave motor.

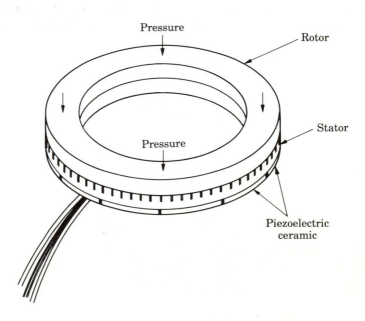

FIG. 1.13.   A ring-shaped travelling-wave motor.

A metal or plastic component is then placed on the metal that is generating travelling waves. This piece becomes the rotor (or slider). If the stator in such a motion and a rigid rotor are in pressure contact with each other, the rotor will be driven by the tangential force at the contact surface resulting from the elliptical movements at the wave crests.

In the rotary motor (Fig. 1.13) two metal rings are pressed together: one carries travelling waves and drives the other. Points on the surface of the former ring move in elliptical trajectories with an amplitude of several micrometres. The latter ring contacts the former at the wave crests. Referring back to Fig. 1.7, notice the grooves on the stator. This is called a comb-tooth ring (or stator). The purpose of the comb teeth is to amplify vibrations. The grooves also allow the dust created by friction to escape and so keep the contact surface dust-free.

1.3.4 The motor's structure

In practice, a thin ring or a lining element is placed between the stator and rotor to increase the friction coefficient and reduce sliding energy loss. The lining material also serves to reduce wear and extends the motor's life span. Although the rotor moves only a micrometre or so in an individual cycle of the elliptical motion, this can add up to a speed of $2\,\mathrm{cm\,s^{-1}}$ at $20\,\mathrm{kHz}$.

Ultrasonic waves are sound waves with frequencies above $20\,\mathrm{kHz}$ and are inaudible to the human ear. The ultrasonic motor, however, does not employ sound waves transmitted through air. Instead, by using piezoelectric ceramic elements, a vibration with frequencies in the ultrasonic range is

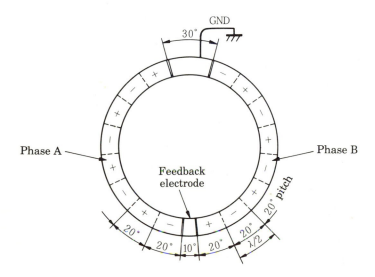

FIG. 1.14. Piezoeletric element: electrode arrangement.

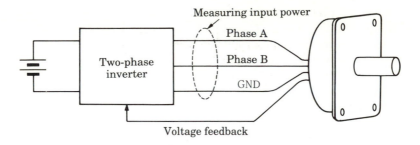

Fɪɢ. 1.15.   Inverter used to create a two-phase a.c. voltage from a d.c. source.

created on the surface of the metal ring. As mentioned earlier, one can reverse the rotation by reversing the voltage phase on piezoelectric element A or B. For instance, one can change the voltage for B to $-C\cos\omega t$. The speed, on the other hand, can be controlled by adjusting the amplitude C.

In the ring-type motor, the electrodes for the piezoelectric elements are arranged as shown in Fig. 1.14. Plus ($+$) and minus ($-$) signs indicate the polarization directions of the ceramic elements. Although this will be discussed at length in Chapter 3, we briefly explain here what this means. When a positive voltage is applied to a segment indicated by a plus sign, it will expand; with a negative voltage, it will contract. The reverse occurs for a segment with a 'minus' polarity. The arrows in Fig. 1.12 are used similarly. Finally, we need a device to convert a d.c. source voltage into a two-phase sinusoidal voltage. This is called an inverter (see Fig. 1.15).

1.4   Ultrasonic linear motors

In 1982, two types of linear ultrasonic motor were constructed and tested in Sashida's laboratory. One of them, shown in Fig. 1.16(a), is the monorail type. The ends of a long metal bar are welded together to form an 'endless' stator rail. A carriage lies on top and is pressed against the rail. This motor is a developed version of the ring-type motor, in which the piezoelectric ceramic vibrators are bonded on to the rail's bottom surface.

The other type, shown in Fig. 1.16(b), is literally a linear motor (with linear movements). Two bolt-tightened Langevin vibrators are installed at the ends of the rail to create travelling waves with large amplitudes: vibrator 1 generates travelling waves, while vibrator 2 absorbs them to prevent reflections. Vibrator 2 is therefore called an absorber. With this arrangement, travelling waves propagate from 1 to 2, but the carriage moves in the opposite direction (i.e. from 2 to 1). One can reverse the carriage movement by reversing the roles of the two vibrators.

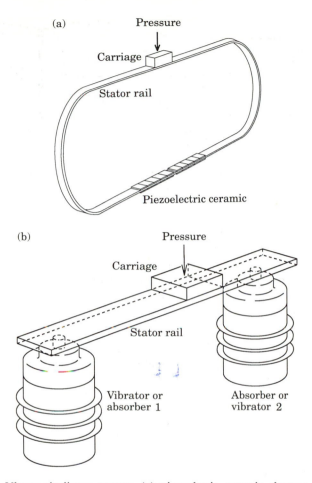

FIG. 1.16. Ultrasonic linear motor: (a) piezoeletric ceramic element attached to rail; (b) using Langevin vibrators.

The carriage can reach a speed of nearly $1 \, \mathrm{m \, s^{-1}}$ in this model. A test apparatus for these two types is shown in Fig. 1.17.

As we see in the above example, the Langevin vibrator converts an a.c. voltage with an ultrasonic frequency into mechanical vibrations of the same frequency, and vice versa (i.e. mechanical vibrations into a.c. voltage).

1.5 Other motor types

So far we have described the basic principles of some representative ultra-sonic motors. After Sashida developed the travelling-wave motor, several

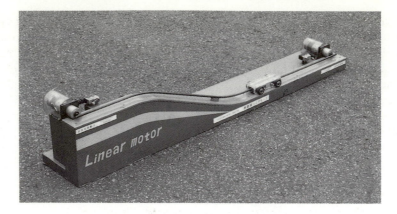

FIG. 1.17.   Experimental linear-motor set-up.

(a)

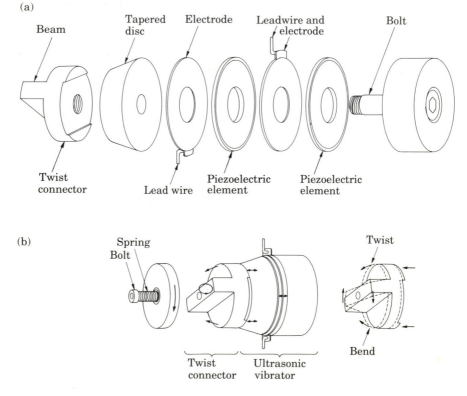

(b)

FIG. 1.18.   Kumada's twist-connector motor: (a) structure of the ultrasonic twist vibrator; (b) mechanism of the twist-vibrator ultrasonic motor.

industry and university research laboratories in Japan actively began research on ultrasonic motors, with the result that to date, several models have been proposed, tested, and improved upon. One research area focuses on the use of multiple vibrational modes. Three examples are given below.

The motor shown in Fig. 1.18 was invented by Kumada, and is called the twist-connector ultrasonic motor. The twist connector is a device positioned between the Langevin vibrator and the rotor. A wide shallow groove is machined on one side, while a beam ridge protrudes from the other. The beam and groove are tilted with respect to each other, and convert axial vibrations into twisting vibrations to create elliptical motion at the beam's surface. The tangential force at the beam's surface then drives the rotor. A complex mechanism is involved which determines the rotational directions of the ellipses; interested readers should consult ref. 3.

The motor shown in Fig. 1.19 was developed by Ueha and colleagues[4], and also combines the use of twisting and axial vibrations. In this motor, the twisting vibrations in the metal block are synchronized with the axial vibrations generated in the three multilayer piezoelectric actuators at the top to create elliptical motion at the surface of the actuators. The reason for the multilayer construction for the axial movement is to obtain enough amplitude when the system is excited at a frequency different from the resonant frequency. The advantage is that the amplitudes in the axial and transverse directions can be controlled independently. The improved model

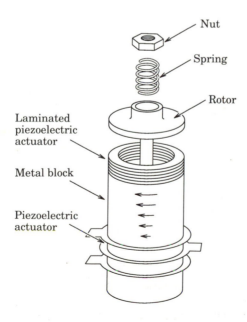

FIG. 1.19. Ueha's multivibrator motor.

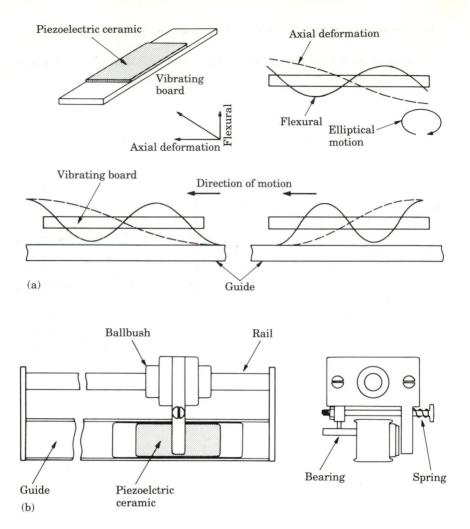

FIG. 1.20.   Tomikawa's motor: (a) bending; (b) axial deformation.

uses one ring-shaped multilayer construction instead of three separate elements.

The motor in Fig. 1.20 is that of Tomikawa[5,6]. It combines flexural waves and axial deformations to create elliptical motion at the edge of the beam. A full treatment of these motors, which employ multiple vibrational modes, is outside the scope of this text.

A number of invaluable attempts at creating new models are reviewed in ref. 7.

1.6 Advantages of the ultrasonic motor and its applications

Ultrasonic motors have several advantages, as illustrated by the following specimen applications.

1.6.1 Little influence by magnetic field

The greatest advantage of the ultrasonic motor is that it neither is affected by nor creates a magnetic field. Regular motors which utilize electro-magnetic induction will not perform normally when subjected to strong external magnetic fields. Since a fluctuation in the magnetic field will always create an electric field (following the principle of electromagnetic induction), one might think that the ultrasonic motor would be affected as well. In practice, however, the effects are negligible. A fluctuation in the magnetic flux density by, say, 1 T (which is a considerable amount) at a frequency of 60 Hz will create an electrical field of $\sim 100\,\mathrm{V\,m^{-1}}$. This is two orders of magnitude below the usual field strength in the piezoelectric ceramic elements and so can be ignored.

This feature is used to advantage in actuators installed in the peripheral equipment of nuclear magnetic resonance units. Nuclear magnetic resonance (NMR) occurs when a nuclear proton resonates with electromagnetic waves

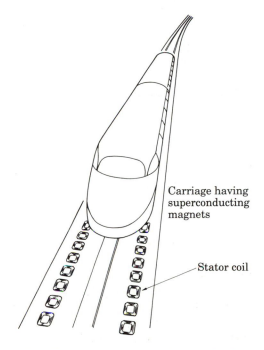

Carriage having superconducting magnets

Stator coil

FIG. 1.21.   Ultrasonic motors used in Maglev linear trains.

at certain frequencies in a strong magnetic field. Using NMR, one can identify an element by measuring its resonant frequency, since this is determined
by the nuclear structure. Besides serving this purpose in physics and chem-
istry, the phenomenon is also used in medical equipment to obtain cross-
sectional images of the brain (nuclear magnetic resonance tomography).

Another use for ultrasonic motors can be found in linear trains using
magnetic levitation (see Fig. 1.21). The coaches are supported by a strong
magnetic field created by superconductivity. Powerful and precision-
controlled actuators are required for this purpose, for which ultrasonic
motors are ideally suited.

1.6.2   Low-speed–high-torque characteristics, compact size, and quiet operation

Ring-type ultrasonic motors can be made very compact in size. Moreover,
as we shall see in Chapters 5 and 8, the travelling-wave motor generates
high torques at low speeds, and no reduction gears are needed. The motor
is also very quiet, since its drive is created by ultrasonic vibrations that are
inaudible to humans. In recent years, machine-generated noise pollution
has become a major problem. In offices, hospitals, hotels, theatres and

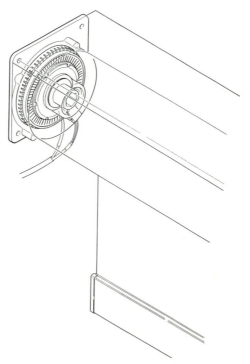

FIG. 1.22.   Ultrasonic motor used for window blind control.

libraries, which require low noise levels, motors that create gear noise can cause problems. In Tokyo's newly built city hall, a thousand ultrasonic motors were installed as automatic control units for window blinds (see Fig. 1.22). The ultrasonic motor was chosen in this case because of its quiet operation.

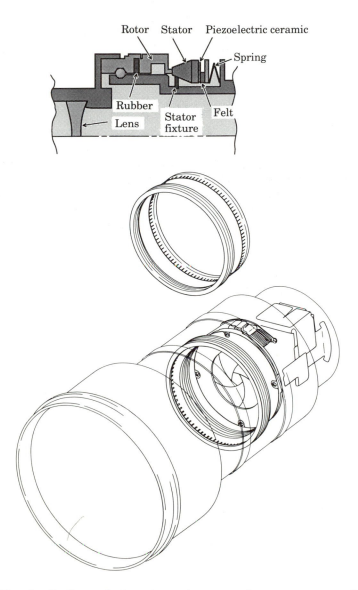

FIG. 1.23. Application of the ultrasonic motor in a camera autofocusing mechanism.

1.6.3 Hollow structure

The drive mechanism of the travelling-wave motor is ring-shaped. Thus, as long as the pressure mechanism between the rotor and stator can be accommodated, the motor can have a shaftless (or hollow) structure. The open space can then be used for various purposes. This feature is utilized in the autofocusing mechanism of some cameras, such as the Canon EOS single-lens reflex camera. As shown in Fig. 1.23, the shaftless motor is installed in the lens frame and reduces the focusing time compared with cameras which use electromagnetic motors inside the camera body. Since reduction gears are not required, this camera is also quieter.

1.6.4 Holding torque and quick response

Because the stator and rotor are pressed together, the ultrasonic motor will maintain its holding torque and act as a brake after being turned off. Owing to the rotor's low inertia, the motor also has a quick response, with a mechanical time constant of 1 ms or so. In other words, it is easy to control.

(1) *X–Y table (Fig. 1.24).* Use of ultrasonic motors on an X–Y table provides a simple servomechanism for positioning control. As soon as the sensor detects its target position and cuts off the power, the motor will stop and hold the table in place. In this, quick response is possible because the braking force exerted by friction is much larger than the inertial force of the rotor.

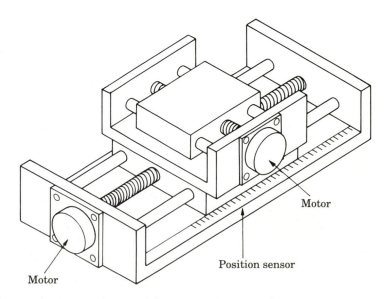

Fig. 1.24. Ultrasonic motor used in coordinate plotter.

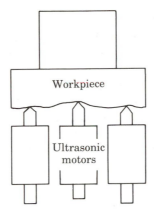

FIG. 1.25.   A workpiece stopper.

(2)   *Workpiece stopper (Fig. 1.25)*. To machine a diecast workpiece, it is usually necessary initially to form a reference plane on the workpiece. With the ultrasonic motor, this process can be eliminated. Several pins driven by ultrasonic motors approach the workpiece when it is placed on the work

FIG. 1.26.   'Maxwell's devil' built for demonstration of quick-response characteristics of the travelling-wave motor: (a) random series of black balls and white ones fall down, gated by a sprocket coupled to a tiny ultrasonic motor at a rate of 100 per second; (b) black balls and white ones are separated by a sensor and another motor.

table. Each pin halts its progress and acts as a fixed stopper as soon as it detects contact with the workpiece via electrical signals. Thus, with multiple pins functioning together as a reference plane, it becomes unnecessary to machine a flat plane.

(3) *Material selector (Fig. 1.26)*. The fast response characteristic of the ultrasonic motor is demonstrated by the machine shown in Fig. 1.26. A motor separates falling black and white balls. After detection of the colour by a photodetector mounted 10 cm above the ultrasonic actuator, only 10 ms is occupied in signal processing and action of the motor.

1.6.5 Compact-sized actuators

The ultrasonic motor's small size and large torque are utilized in several applications. The motor shown in Fig. 1.27 is installed inside a narrow pipe

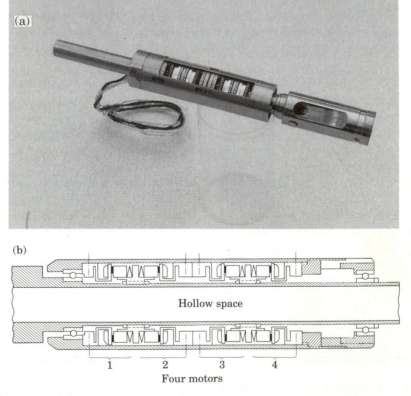

FIG. 1.27. Ultrasonic motor in a robot actuator and its specifications (a) Inside view and welding head (a lens and a mirror are mounted on the side hole); (b) cross-section. Specifications: weight, 30 g; outer diameter, 15 mm; inner diameter, 7 mm; length, 56 mm; starting torque, 0.15 N m; input power, 5 W.

and has a hollow structure to allow room for cables. A high torque is required for its intended application in robotics. At the design specifications given in the figure, this 30 g motor, which has a four-tiered structure, has a starting torque of 0.15 N m; this kind of performance is difficult to achieve with an electromagnetic motor. The ultrasonic motor's hollow structure is necessary for this application also. It would be difficult to design a similar device using an electromagnetic motor with its gear head and still satisfy the required specifications.

1.7   Practical limitations in piezoelectric ceramics; future research areas

Like any other technology, the ultrasonic motor has shortcomings as well as advantages and there is ample room for further research to overcome these shortcomings and improve the motor. In this section, we briefly discuss some of these issues.

1.7.1   Limitations of the ultrasonic motor

The ultrasonic motor is subject to limitations of the piezoelectric ceramic. The motor's vibrations create alternating stresses in the ceramic elements which can result in fatigue failure. The ceramic is strong in compression but weak in tension, with respective failure limits in the ratio of 30 to 1. In the Langevin vibrator, tensile failure is prevented by the application of a compressive prestress applied with a bolt. In the travelling-wave ring motor, large amplitudes are not possible, since the compressive and tensile stresses generated in the ceramic pieces are of equal magnitude.

Above a certain temperature, known as the Curie point, the ceramic loses its piezoelectric properties. The motor, however, is unlikely to reach the Curie point temperature, which is $>300°C$. Instead, weakening of the adhesive bond will occur below the Curie point. The Young's modulus of the metal will also change with temperature, causing a change in the resonant frequency. This will lower the motor's efficiency. Since it is not possible to control the Young's modulus, in practice the excitation frequency is adjusted to match the resonant frequency.

1.7.2   Basic research is needed to improve efficiency

Although the wedge-type motor has a high efficiency, at the same time it has a short life span owing to rapid wear. The Langevin vibrators amplify the ceramic's vibrations of $<0.1 \mu m$ to several micrometres, which is further amplified by the horn, accounting for the high efficiency. On the other hand, the Langevin vibrator wears rapidly owing to repeated collisions with the rotor.

In the travelling-wave motor, there is less wear since the drive (linear or rotary) is created by flexural waves propagating in a beam and no collisions take place. However, frictional losses are large owing to the complex nature of the vibrations, and efficiencies are at most ~50 per cent at the present time. As we shall see in Chapter 5, if sliding and deformation losses at the stator–rotor interface can be eliminated, the efficiency will be further increased.

One factor that keeps the ring-type motor's efficiency low is the very small amplitudes of the flexural waves created in the stator. In order to improve the efficiency, an in-depth study of contact surface behaviour is needed.

References

1. Barth, H. V. (1973). Ultrasonic driven motor. *IBM Technical Disclosure Bulletin*, **16**, 2263.
2. Gromakovskii, V. A. *et al.* (1978). On the possibility of using a piezoelectric motor for direct actuation of the drive shaft of a video tape recorder. *Tekhnika Kino i Televideniya*, (5), 33–43.
3. Kumada, A. (1985). A piezoelectric motor. *Japanese Journal of Applied Physics*, **24**, Supplement 24-2, 739–741.
4. Kurosawa, M. and Ueha, S. (1987). An ultrasonic motor using a vibrator and multilayered piezoelectric actuators. Japanese Institute of Electrocommunication Reports, US87-31, pp. 27–32.
5. Tomikawa, Y., Ogasawara, T., and Takano, T. (1989). Ultrasonic motors contructions/characteristics/applications. *Ferroelectronics*, **91**, 163–78.
6. Takano, T. and Tomikawa, Y. (1989). Linearly moving ultrasonic motor using a multi-mode vibrator. *Japanese Journal of Applied Physics*, **28**, Supplement 28-1, 164–6.
7. Tomikawa, Y. and Ueha, S. (1991). *Ultrasonic motors*, (new edn). Ch. 4. TRICEPS, Tokyo.

2. Theoretical treatment of component elements

We presented a general outline of the ultrasonic motor in the previous chapter. In this chapter, we shall study the basic principles governing the action of each component of a motor to develop an overall theory of the ultrasonic motor in later chapters.

2.1 Mechanical friction

In a social context, 'friction' denotes something undesirable, such as 'trade friction' or 'friction among people'. However, mechanical friction is not only desirable but indispensable to us. For instance, friction between rubber tyres and the road surface allows an automobile to run. Similarly, friction between steel surfaces enables a train to travel fast. If friction and gravity did not exist, we would not be able to walk.

Friction also plays a vital role in the ultrasonic motor. Hence we shall first examine this phenomenon. One needs to push a boat floating on water only slightly to make it move, whereas a considerable effort is required to move a boat resting on a sandy beach. As illustrated by this example, when a force is applied to an object that rests on a solid surface, a resisting force is created at the contact surface. This resistance is caused by microscopic irregularities on the two surfaces which 'mesh' with each other when in contact. This resisting force is known as friction, and its magnitude depends on such factors as the material, surface roughness, wetness, and the presence of lubricants, as well as the force pressing the two objects together. In general, there are two types of friction: static and kinetic.

2.1.1 Static friction

An object with weight W rests on a surface and a horizontal force F is applied, as shown in Fig. 2.1. For small values of F, the object remains still because an opposing force f of equal magnitude is automatically generated. This resisting force f at the contact surface is called the static frictional force. As F is increased, f will also increase. If F exceeds a certain value, the object will start to move; the frictional force at this moment is called the maximum static frictional force. If F_0 is the maximum static frictional force and W is the normal force acting on the contact surface, then F_0 is proportional to W:

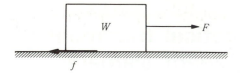

FIG. 2.1.   Resisting force f acting on the contact surface.

$$F_0 = \mu_s W. \qquad (2.1a)$$

The proportionality constant μ_s is known as the coefficient of static friction. This relationship is an empirical formula which is not affected by the surface area as long as the pressure is not excessively large. The coefficient depends on such factors as surface roughness, presence of lubricants, and temperature. Table 2.1 shows the coefficient of static friction for a few materials. However, it is not certain that these values can be applied directly in the case of the ultrasonic motor.

2.1.2   Kinetic friction

The body in Fig. 2.1 will start to slide when F overcomes the maximum static friction. When the body is moving, the resistance to the applied force is less than the maximum static friction. This resistance is known as the kinetic friction. The kinetic friction is proportional to the normal force acting on the contact surface:

$$F' = \mu_m W \qquad (2.1b)$$

where F' is the kinetic friction and μ_m is known as the coefficient of kinetic friction.

2.1.3   Microscopic examination of friction

We defined the maximum static friction as the resisting force acting on a body immediately before it starts to move. By 'move' we mean the apparent movement observed with the naked eye. In the ultrasonic motor, however,

Table 2.1. Coefficients of static and kinetic friction

I	II	Static	Kinetic
Steel	Steel	0.7	0.5
Bronze	Steel	0.51	0.44
Copper	Steel	0.53	0.36

Note: body I rests on surface II.

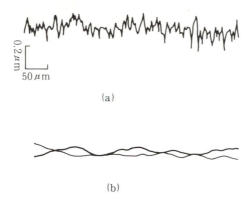

(a)

(b)

Fig. 2.2. Contact between finished surfaces and rough surfaces: (a) finished surface; (b) contact between surfaces.

it is necessary to examine the microscopic behaviour, since the vibrations which occur are on the order of micrometres. Finished surfaces produced by machining, grinding or polishing are still 'rough' when viewed microscopically (Fig. 2.2a). When these surfaces are pressed against each other, the true contact area is said to be from 1/1000th to 1/10 000th of the apparent area observed by the naked eye (Fig. 2.2b).

By setting up a model contact point as shown in Fig. 2.3, Kato[1] obtained the relation between displacement x and F/W for platinum-on-platinum and steel-on-steel contact. The results are shown in Fig. 2.4. Notice that a displacement occurs even if the applied force is minute. What was defined as the coefficient of static friction above is the F/W value for higher x values. Strictly speaking, therefore, μ_s for zero displacement does not really exist.

Kato also performed an experiment on kinetic friction. As shown in Fig. 2.5(a), a body M rests on a plane and the spring connecting M to the fixed wall is stretched to its maximum limit. As soon as the plane starts to move to the left at velocity v, as in (b), M will start sliding towards the wall owing to the spring's pull. Until a steady state is achieved at $x = x_s$,

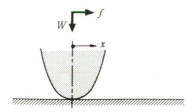

Fig. 2.3. Experimental model for friction (from ref. 1): sliding at the contact point.

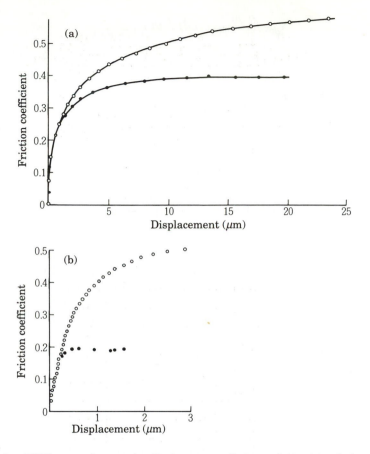

FIG. 2.4. *F/W* vs. microscopic displacement (from ref. 1): (a) platinum on platinum; (b) steel on steel. (●) Lubricated, (○) non-lubricated surfaces.

the body's relative velocity is greater than v, as shown in (c). The coefficient of friction μ varies during this transitional period; see (d).

2.2 Hysteresis

In describing static or kinetic friction above, we assumed that the magnitude of the applied force was gradually increased, and its direction remained the same. What will the force–displacement relationship be if the force is applied in a more complex manner?

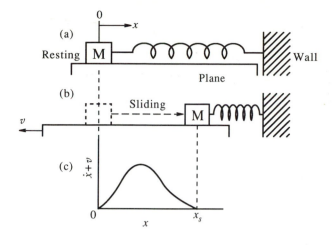

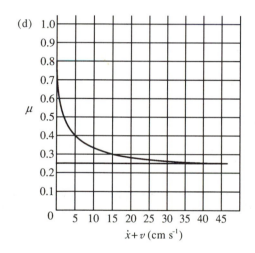

FIG. 2.5. Experiment for observing transition from static to kinetic friction (see text).

2.2.1 *Hysteresis in force–displacement relationship*

Figure 2.6 shows the relationship between torque and steering angle for an automobile's steering mechanism. Figure 2.6(a) shows the torque–angle hysteresis loops when the steering wheel is slowly rotated first in the clockwise direction, then in the counterclockwise direction. Figure 2.6(b) shows minor loops which appear when several 'backsteps' are performed during the manoeuvre. As we shall see later, the minor hysteresis loops can be interpreted as the result of a combined friction mechanism.

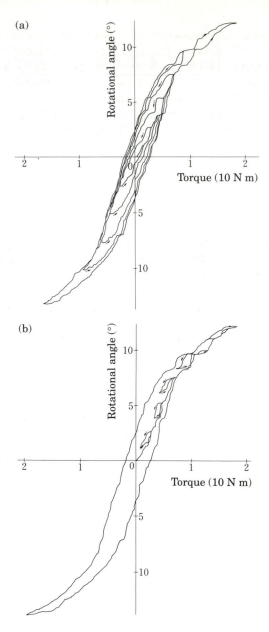

Fig. 2.6. Hysteresis in an automobile's steering mechanism: (a) major loop gradually extended; (b) minor loops on the initial curve portion.

The reader may rightly question our discussion of large steering angles after mentioning microscopic displacements above, since these two phenomena belong to quite different dimensional scales. Yet it is not unreasonable to assume that the microscopic and macroscopic force–displacement relationships exhibit similar hysteresis phenomena. Various types of ultrasonic motor have been developed so far and will continue to be developed in the future. The diverse wave phenomena utilized in these motors will result in as many types of friction behaviour. It is thus important to construct a theoretical framework that can be used to describe all these complex frictional actions.

2.2.2 A simple model

Figure 2.7 shows a body M resting on a rough surface with pressure applied from above (or self-weight). Body M is connected via a spring to a fixed wall. We shall consider a horizontal force f acting on the body (a force acting towards the right is considered as positive). The body's initial position $x = 0$ corresponds to the spring's undeformed state. Because of friction, the body will remain motionless until f reaches $f_s = \mu_s W$. A displacement x occurs as soon as f exceeds f_s (see Fig. 2.8). This can be expressed in

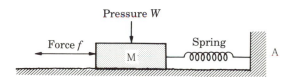

FIG. 2.7.   A force f is applied to an object M resting on a rough surface.

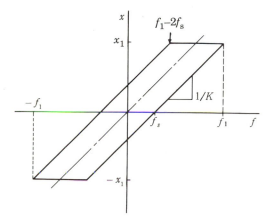

FIG. 2.8.   Simple parallelogram hysteresis.

terms of the spring constant K as

$$x = (1/K)(f - f_s). \qquad (2.2)$$

For simplicity, we shall assume that the coefficients of static and kinetic friction are equal. Then f is increased to f_1, which corresponds to $x = x_1$. After this, f is decreased, but owing to friction the body will not move until $f = f_1 - 2f_s$. The displacement decreases as f is further decreased to zero, then increased in the reverse (or negative) direction, which corresponds to the line with slope $1/K$. At $f = -f_1$, $x = -x_1$. If f is once again increased in the positive direction, M will not move until $f = -(f_1 - 2f_s)$. As f continues to increase, displacement x will increase following the line with slope $1/K$ and the loop is completed. A major hysteresis loop is formed by this process.

A minor loop can be obtained by decreasing f before f_1 is reached. If f is decreased by an amount less than $2f_s$, a line will form instead of a loop (ab in Fig. 2.9). If the decrease exceeds $2f_s$, the path will follow a loop as shown (defc).

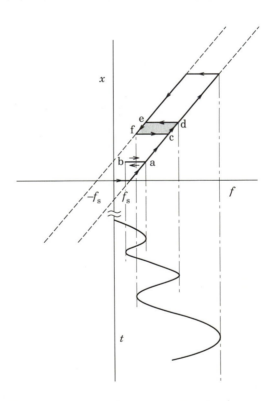

FIG. 2.9.   Two types of minor loop appearing in the simplest model.

2.2.3 System with two springs: minor hysteresis loops with finite areas

In Fig. 2.10, the force f acting on body M_2 acts on M_1 at the same time. Therefore, movement will occur at the contact surface with lower friction. Friction between the base and M_1 with spring K_1 is given by the hysteresis parallelogram of element 1 in Fig. 2.11. Similarly, hysteresis for friction between M_1 and M_2 with spring K_2 is given by element 2. The f–x plot for the combined system is obtained by adding the respective displacements, keeping a common f-axis. We can see that the minor loop in the combined hysteresis diagram has a finite area. The system shown in Fig. 2.12 consists of three bodies. Springs connect M_2 and M_3, but no static friction exists between the two bodies. The frictional hysteresis loop for this system is shown in Fig. 2.13.

2.2.4 A model for saturation

Saturation displayed in a complex hysteresis loop indicates non-linear behaviour in the springs. Figure 2.14 shows two systems displaying saturation caused by springs being stretched or compressed to their respective limits.

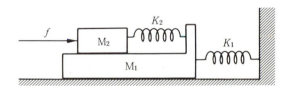

FIG. 2.10. Two springs with two contact surfaces.

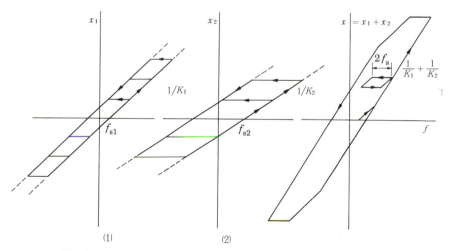

FIG. 2.11. Formation of a minor loop in a two-body model.

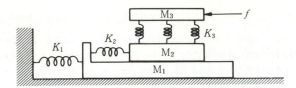

FIG. 2.12. A system with three bodies M_1, and M_2 and M_3.

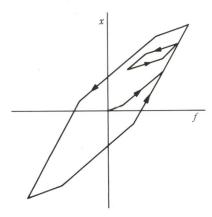

FIG. 2.13. Hysteresis loop for system with three bodies, three spring connections, and two frictional interfaces.

2.3 Equivalent circuits for mechanical systems

It is often useful to represent a problem in mechanics by an equivalent circuit. In Chapter 6 we shall bring together concepts developed in the first five chapters to construct an equivalent circuit for the ultrasonic motor. With this circuit, we can then calculate the motor's characteristics. As a first step, we shall present the basic idea of the equivalent circuit to determine the static and dynamic behaviour in force and velocity transmission of a system where friction plays an essential role.

2.3.1 Mechanical circuit

Figure 2.15(a) shows a mechanical circuit in which a force f causes a velocity v. The applied force f is in equilibrium with the sum of the maximum static friction F and the spring's reaction Kx:

$$f = F + Kx \qquad (2.3)$$

The body's displacement x and velocity v have the following relationship:

$$x = \int_0^t v \, dt. \qquad (2.4)$$

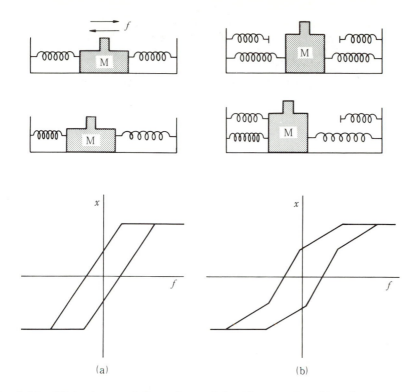

FIG. 2.14.   Major hysteresis loops for models with saturation: (a) spring has stretch and compression limits; (b) in addition to saturation, spring constant is a function of x.

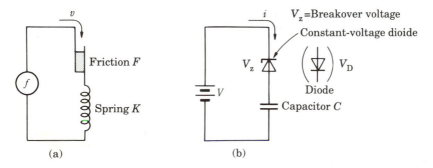

FIG. 2.15.   Corresponding mechanical and electrical circuits: (a) mechanical; (b) electrical.

This circuit is analogous to an electrical circuit with a diode and capacitor in series, shown in Fig. 2.15(b). In the mechanical circuit in Fig. 2.15(a), the object will remain motionless as long as f is below the maximum static friction F. When f exceeds F, a displacement will take place and f equals the sum of F and Kx. In Fig. 2.15(b), no current flows if the source voltage remains below the breakover voltage of the constant-voltage diode. As soon as the source exceeds this voltage, a current i flows and as a result a charge q is stored in the capacitor. $V - V_z$ is then in balance with q/C. Or,

$$V = V_z + \frac{1}{C} q. \tag{2.5}$$

The charge and current have the following relationship.

$$q = \int_0^t i \, dt. \tag{2.6}$$

Several types of diode exist. Figure 2.16 shows three such types. The one in (a) is a regular diode which prevents current from flowing in the reverse direction, but allows current in the forward direction when the applied voltage exceeds a certain value, say 0.6 V. Fig. 2.16(b) shows a constant-voltage diode: in the forward direction it behaves as a regular diode; in the reverse direction, it blocks any current as long as the applied voltage is below the breakover voltage V_z. Any amount of current is allowed to flow once V_z is exceeded, while the terminal voltage remains constant at V_z. The trigger diode in (c) resembles friction in behaviour. The breakover voltage corresponds to the maximum static friction. When the applied voltage exceeds this value, current will flow in the circuit and at the same time the terminal voltage will drop. This is similar to friction, where the kinetic friction is lower than static friction (i.e. $\mu_m < \mu_s$).

In reality, mechanical friction is a fairly complex phenomenon. However, we shall assume that it can be represented by an appropriate diode in the equivalent circuit.

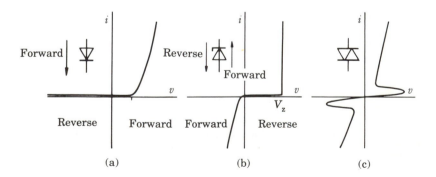

FIG. 2.16.   Three types of diode: (a) regular; (b) constant-voltage; (c) trigger.

2.3.2   Systems with multiple bodies

The mechanical system of Fig. 2.10 can be represented by a parallel circuit of two branches, with each branch containing a diode and capacitor connected in series. For the system in Fig. 2.12, three branches are connected in parallel, although in one of them $F_3 = 0$ (i.e. no diode). In general, a system in which friction exists between each component can be represented by a circuit of parallel elements as in Fig. 2.17.

If the friction exists between a fixed surface and the bodies M_1 and M_2, as shown in Fig. 2.18(a), the circuit would be as in (b). This is explained as follows. M_1 will not move as long as the applied force f remains below friction F_1 (i.e. no current will flow as long as the voltage is lower than the breakover voltage). When f exceeds F_1 by a small amount, M_1 will move some distance to compress spring K_1, although the force is not yet large enough to displace M_2. (In the equivalent electrical circuit, a charge is stored in C_1 and a terminal voltage is created. However, if this voltage

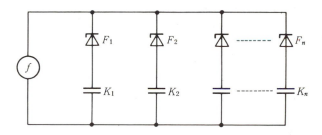

FIG. 2.17.   Circuit with parallel elements when friction acts between bodies.

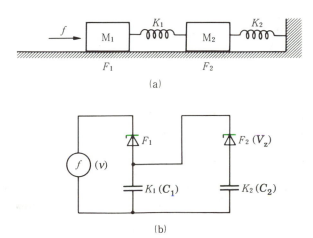

FIG. 2.18.   (a) Two bodies resting on a fixed surface; (b) the equivalent circuit.

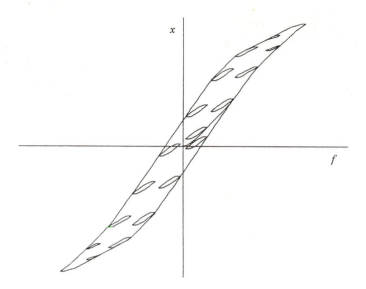

FIG. 2.19.   Hysteresis loop obtained with desktop computer.

remains below V_z, no charge will be stored in C_2.) As f is increased, spring K_1 contracts further. When $K_1 x$ exceeds F_2, body M_2 will be displaced also. (The behaviour of the equivalent circuit should be obvious to the reader.) The hysteresis loop in Fig. 2.19 was obtained using a desktop computer: it represents a circuit with five parallel branches and contains saturation.

2.3.3 Dynamic characteristics

So far, we have presented a static behaviour analysis of a system; that is, the outcome of some process was obtained, leaving the transitional behaviour unknown. Our next step is to extend our theory so that we can examine the dynamic behaviour of a system. The equation of motion for the mechanical system shown in Fig. 2.20(a) is given by

$$M \frac{dv}{dt} + Dv + K \int_0^t v\,dt + F = f(t). \tag{2.7}$$

This system can be represented by the mechanical circuit in (b). Here, the mass M and viscous dashpot D can be represented by an inductance and a resistance respectively. This mechanical circuit corresponds exactly with the electrical circuit in (c), the equation for which is

$$L \frac{di}{dt} + Ri + \frac{1}{C} \int_0^t i\,dt + e_D = V(t). \tag{2.8}$$

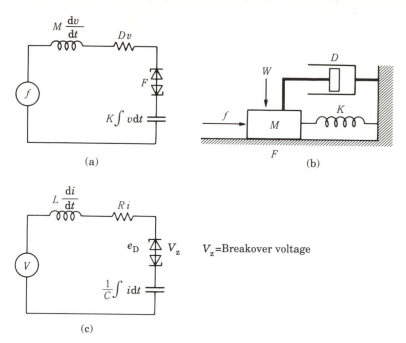

FIG. 2.20. Mechanical circuit and equivalent electrical circuit for dynamic analysis: (a) mechanical circuit; (b) mechanical model; (c) electrical circuit.

Note that F and e_D are not set constants but vary according to the situation. Consider the electrical circuit, with the constant voltage diode acting in reverse bias:

(a) When $|v_D| \leq V_z$ (breakover voltage), the diode prevents any current flow, and the reverse electromotive force e_D created by the diode equals the applied voltage:

$$e_D = v_D. \tag{2.9}$$

(b) When $|v_D| > V_z$, current flows through the diode.
  (i) If $v_D > 0$,

$$e_D = V_z. \tag{2.10}$$

  (ii) If $v_D < 0$,

$$e_D = -V_z. \tag{2.11}$$

2.3.4 Dynamic behaviour of multibody systems

Figure 2.21 shows two cases of two connected bodies of masses M_1 and M_2. From the figure, we can deduce the following.

(1) Forces necessary to accelerate the bodies are given by $M_1(dx_1/dt)$ and $M_2(dx_2/dt)$, where x_1 and x_2 are the absolute velocities for bodies M_1 and M_2.

(2) The spring's reaction f_K is determined by the relative displacement between the two bodies as

$$f_K = K_1(x_1 - x_2).\tag{2.12}$$

(3) Friction (static or kinetic) is determined by the relative velocity $\dot{x}_1 - \dot{x}_2$ for Fig. 2.21(a), and by the absolute motion (or velocity) for (b).

Thus we obtain the following equations of motion.
    For the system in Fig. 2.21(a),

$$M_1\frac{d\dot{x}_1}{dt} + D_1(\dot{x}_1 - \dot{x}_2) + F_1 + K_1(x_1 - x_2) = f\tag{2.13}$$

$$M_2\frac{d\dot{x}_2}{dt} + D_2\dot{x}_2 + F_2 + K_2x_2 = f - M_1\frac{d\dot{x}_1}{dt}\tag{2.14}$$

where F_1 is determined by the relative velocity.
    For the system in Fig. 2.21(b),

$$M_1\frac{d\dot{x}_1}{dt} + D_1\dot{x}_1 + F_1 + K_1(x_1 - x_2) = f\tag{2.15}$$

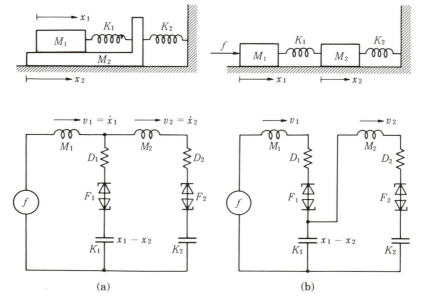

(a)   (b)

FIG. 2.21. Two-body systems: (a) behaviour determined by relative velocity $\dot{x}_1 - \dot{x}_2$; (b) friction for M_1 and M_2 determined by absolute velocities.

$$M_2 \frac{d\dot{x}_2}{dt} + D_2 \dot{x}_2 + F_2 + K_2 x_2 = K_1 (x_1 - x_2) \qquad (2.16)$$

where F_1 and F_2 depend on the respective velocities, $\dot{x}_1$ and $\dot{x}_2$, of the bodies.

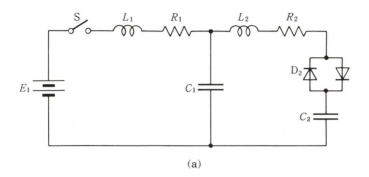

(a)

```
100 CLS
110 L1=.01: L2=.005 : R1=1 : R2=1.1
120 C1=.00001: C2=.00002: D2=.6
130 E=10
140 VER=80: HOR=300
150 LINE (VER,HOR-200)-(VER,HOR)
160 LINE (VER+400,HOR)-(VER,HOR)
170 FOR M=1 TO 20
180 LINE (VER+M*20,HOR)-(VER+M*20,HOR+5)
190 NEXT M
200 LINE (VER,HOR-200)-(VER,HOR)
210 LINE (VER+400,HOR)-(VER,HOR)
220 FOR M=1 TO 20
230 LINE (VER,-M*10+HOR)-(VER-5,-M*10+HOR)
240 NEXT M
250 LOCATE 3 ,7:  PRINT "UNIT"
260 LOCATE 3 ,8:  PRINT "Volt"
270 LOCATE 30,20: PRINT "UNIT"
280 LOCATE 35,20: PRINT "ms"
290 LINE (VER,HOR)-(VER,HOR)
300 '*** COMPUTE ***
310   DT=.00002
320   FOR N=1 TO .02/DT
330   '*** INTEGRATE ***
340   INTEGRAL1=INTEGRAL1+(E-(Q1-Q2)/C1)*DT
350   DQ1=1/L1*(INTEGRAL1-R1*Q1)*DT: Q1=Q1+DQ1
360   VOLT=(Q1-Q2)/C1-Q2/C2
370   IF DQ2 >0 THEN VD=D2: GOTO 380: ELSE VD=-D2
380   DQ2=1/L2*(INTEGRAL2-R2*Q2)*DT: Q2=Q2+DQ2
390   INTEGRAL2=INTEGRAL2+((Q1-Q2)/C1-Q2/C2-VD)*DT
400   VC1=(Q1-Q2)/C1: VC2=Q2/C2
410   LINE -(VER+20000!*DT*N,HOR-10*(VC2))
420 NEXT N
430 END
```

(b)

Fig. 2.22.   (a) Sample electrical circuit; (b) BASIC program.

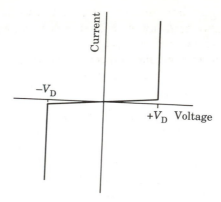

FIG. 2.23. Diode characteristics.

Although analysing the behaviour of a multibody system can be quite cumbersome when several types of frictional behaviour are involved, in principle one can solve the relevant differential equations in a fairly straightforward manner using a desktop computer, once the frictional elements are identified properly. This approach is valid regardless of the number of elements. An example is given in Fig. 2.22. The BASIC program (shown in (b)) determines the terminal voltage on C_2 after switch S is closed in the circuit in (a). The diode's characteristics are shown in Fig. 2.23: a very small amount of current is able to flow in the voltage range between $\pm V_D$. The program result is shown in Fig. 2.24. In the program, DT, as in line 260, represents the time increment used for integration.

2.4   Remarks on equivalent-circuit applications

Although our immediate motive for introducing the equivalent-circuit method in this chapter was to evaluate the dynamic aspects of a system's

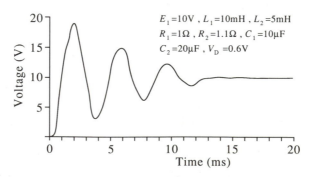

FIG. 2.24.   Program result of Fig. 2.22(b).

frictional behaviour, we have another goal from a wider perspective. It is to enable us, after developing an equivalent circuit for an piezoelectric element and Langevin vibrator in the next chapter, to bring all these elements together to construct an equivalent circuit so as to analyse the motor with a load in Chapter 6. Below we discuss a few useful rules that apply when constructing an equivalent circuit.

(1) *If a constraint exists between two elements, connect in series.* When rotary bodies J_1, J_2, and J_3 are mechanically linked with gears as in Fig. 2.25(a), the speeds and rotational angles of the three bodies are interdependent. In such a case, the three elements should be connected in series. The angle of rotation or angular velocity of one body is sufficient to describe the system. The equivalent spring constant for the equivalent circuit can be determined by the gear ratios, as shown in (b). In this case, it is assumed that there is no backlash in the gears and furthermore that any torque loss occurs due to the friction F or viscosity D.

(2) *If no constraint acts between two bodies, connect them in parallel.* In Fig. 2.26(a), the two bodies M_1 and M_2 behave independently of each other, with force f distributed between the two. This can be represented by the parallel circuit elements in (b). Although the mechanism shown in (c) is slightly more complex, it can be expressed by a similar circuit. Owing to the torque applied to rod J_1, rod J_2 will either rotate about its own axis or revolve around J_3: if J_3's movement is restrained, J_2 revolves around J_3 causing spring K_1 to deform; on the other hand, if spring K_1's movement is restrained, J_3 rotates via the bevel gear mechanism, deforming spring

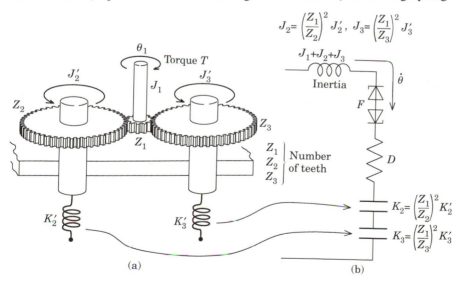

$$J_2 = \left(\frac{Z_1}{Z_2}\right)^2 J_2', \quad J_3 = \left(\frac{Z_1}{Z_3}\right)^2 J_3'$$

Fig. 2.25. Torque T acting on J_1 will also rotate J_2 and J_3. This can be represented by elements in series.

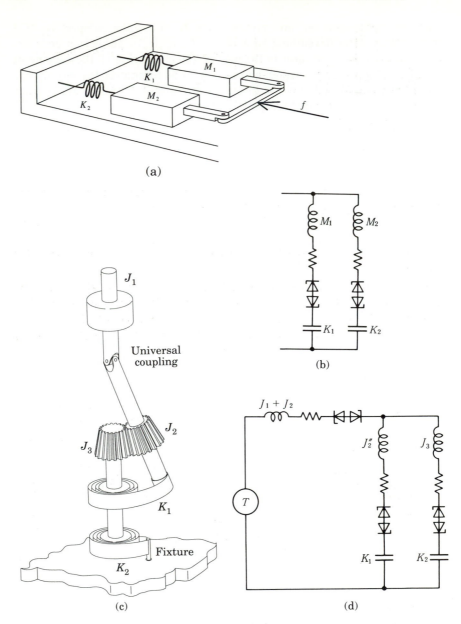

FIG. 2.26.   System with no constraints between bodies: (a) force acts on each body; (b) equivalent circuit for (a); (c) J_2 is able to rotate about its own axis or J_3's axis; (d) equivalent circuit for (c).

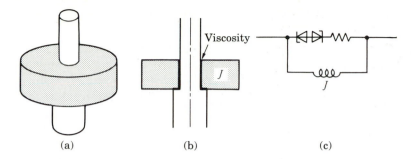

Fig. 2.27. Viscous–inertial damper and its equivalent circuit element: (a) external view of damper; (b) cross-section; (c) equivalent circuit.

K_2. In the figure, J_2'' represents the moment of inertia of J_2 with respect to J_3's axis.

(3) *Viscous–inertial damper*. In Fig. 2.27(a), the ring rotates about the shaft freely, but a highly viscous grease fills the gap. High-frequency vibrations applied to the shaft will not be transmitted to the ring, which has a large inertia. The vibration is thus 'filtered out' by the resistance of the viscous

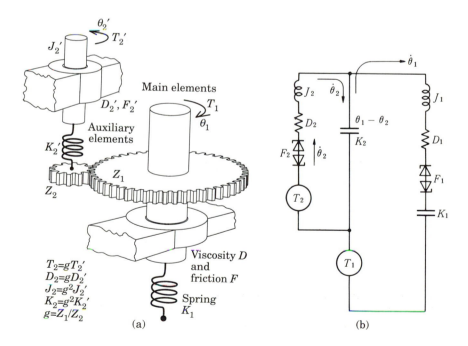

Fig. 2.28. Auxiliary torque T_2 acts through spring K_2. J_1 represents the inertia of the assembly of the two gears and main shaft. (a) Structure; (b) equivalent circuit.

fluid. However, the ring will follow slow movements of the shaft. The equivalent circuit element is shown in (c).

(4) *When an auxiliary torque is applied.* We shall consider the complex mechanism shown in Fig. 2.28. The main drive to spring K_1 is supplied by torque T_1. An auxiliary torque T_2' is provided via spring K_2' and a gear. The equivalent circuit is given in (b). This can be interpreted as follows. The sources of torque act on J_1: the main torque T_1 and the torque acting on the gear via spring K_2. The latter is given by $K_2(\theta_1 - \theta_2)$, and is equal to T_2 less the sum of (a) the torque necessary to accelerate J_2, and (b) the friction. Note that the symbols marked with a prime ($'$) in (a) indicate the actual values for elements in the auxiliary drive. In (b) these have been converted to equivalent values with respect to the main system and hence are not marked so.

2.5   Treatment of loads

So far, our discussion has mostly been about the motor's mechanism up to the stage of producing torque at the shaft. To conclude this chapter, we shall touch upon the subject of loads represented in equivalent circuits. A

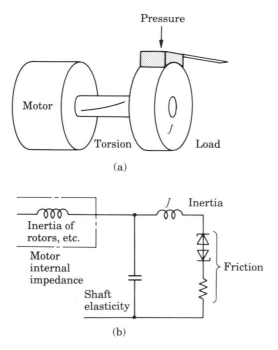

Fig. 2.29.   Treatment of motor load: (a) motor drives an inertial and a frictional load; (b) equivalent circuit.

simple load consists of inertia and friction, as shown in Fig. 2.29(a). The shaft's elasticity is represented by an equivalent capacitance, which is placed in parallel with the load impedance, then connected to the motor's internal impedance.

Reference

1. Kato, K. (1989). Friction between a rotor and stator. In *Ultrasonic electronics: new applications of piezoelectricity*, 25th Tohoku University Communications Research Symposium, III–5, pp. 101–6.

3. The piezoelectric element and vibrator

The conversion of electrical energy into mechanical oscillations which takes place in the piezoelectric element can be separated into two areas of study. One is the study of piezoelectric material and its properties. The other is the mathematical treatment of the electrical-to-mechanical energy transformation. There is quite an extensive literature covering both of these aspects in detail: ref 1 and 2, for example, deal with piezoelectric materials, while ref 3 and 4 discuss the electromechanical energy transformation.

We restrict our presentation to those topics which are necessary for our treatment of the ultrasonic motor, mostly summarizing the results achieved by early researchers. With regard to the treatment of the electrical-to-mechanical energy transformation, however, we shall make use of the equivalent circuit method, and relate it to the subject of friction as treated in Chapter 2.

3.1 The piezoelectric effect

The piezoelectric effect in quartz was discovered in 1880 by the brothers J. Curie and P. Curie. When certain types of crystals are subjected to tensile or compressive forces (or stresses), the resulting strain causes a polarized state in the crystal, and an electric field is created. Conversely, if a crystal is polarized by an electric field, strains along with resulting stresses are created. Together, these two effects are known as the piezoelectric effect. The two aspects are sometimes distinguished as the positive and reverse effects.

In crystals which show piezoelectric properties, mechanical quantities such as stress (T) or strain (S), and electrical quantities such as electric field (E), electric displacement (flux density) or polarization (P), are interrelated. This phenomenon is called electromechanical coupling.

3.1.1 Longitudinal and transverse effects

Among various piezoelectric phenomena, the longitudinal and transverse effects are particularly important. In the longitudinal effect, deformations take place parallel to the electric axis as in Fig. 3.1(a); in the transverse effect, deformations occur at right angles to the electric axis, as in (b). In

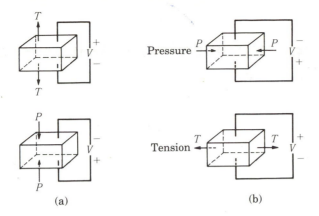

Fig. 3.1.   The piezoelectric effect: (a) longitudinal effect; (b) transverse effect.

practice these two types of effect take place at the same time. Moreover, a third torsional effect is also known.

3.1.2   Piezoelectric ceramics

Quartz and barium titanate are some of the materials which display the piezoelectric effect. Tanaka *et al.*[1], in collaboration with Murata Mfg Co. Ltd, announced the Langevin vibrator in 1951. Under the product name of PZT, Clevite Inc. has produced a material which exhibits even stronger piezoelectric properties. This is a solid solution composed of $PbZrO_3$–$PbTiO_3$. Currently, the term PZT denotes all materials that are created (by metamorphosis) from $Pb(Zr–Ti)O_3$. PLZT denotes materials whose main compositions are $(Pb,La)(Zr,Ti)O_3$.

3.1.3   The piezoelectric phenomenon

From a microscopic viewpoint, two types of strain can be observed in a crystal placed in an electric field: one is proportional to the field strength; the other is proportional to the square of the field strength. The former is the piezoelectric effect in its stricter definition, while the latter is some-times distinguished as the electrostrictive phenomenon. Practical piezoelec-tric ceramics have a complex multidomain structure under the microscope and exhibit quite complex behaviour. Figure 3.2, for example, plots strains in the direction of the applied electric field (i.e. longitudinal effect) for a PLZT material. The state of the material is determined by its previous history, and this property is referred to as hysteresis, similar to the mech-anical hysteresis seen in the preceding chapter. As the hysteresis loop shown here is quite complex, its explanation requires new terminology and a model

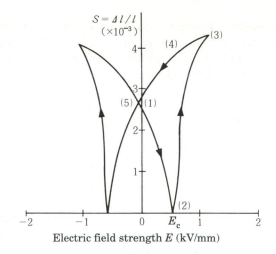

FIG. 3.2. Dielectric strain characteristics for a PLZT piezoelectric ceramic (from ref. 2).

shown in Fig. 3.3, which illustrates how so-called 'poling' is undertaken on a multidomain piezoelectric material.

3.1.4   Terminology

(1)   *Polarization.* Polarization, which is usually denoted by P, is related to electric displacement (or electric flux density) D through the linear expression

$$D_i = P_i + \varepsilon_0 E_i \tag{3.1}$$

where the subscript i represents any of the three coordinates x, y, and z, and ε_0 is the permittivity of free space, equal to $8.854 \times 10^{-12}\,\mathrm{F\,m^{-1}}$.

In most piezoelectric materials D and P are non-linear functions of E and may depend on the previous history of the material. When the term $\varepsilon_0 E$ in the above expression is negligible compared with P (as in most cases), D is nearly equal to P.

(2)   *Permittivity.* This parameter, which is denoted by ε, is defined as the incremental change in electric displacement per unit electric field when the magnitude of the measuring field is very small compared with the coercive electric field, which is the electric field denoted by E_c in Fig. 3.2.

(3)   *Remanent polarization.* The value of the polarization that remains after an applied electric field is removed is defined as remanent polarization.

(4)   *Poling and switching.* The crystal in our model has a polycrystalline structure consisting of many domains. Poling is the process by which a d.c.

electric field exceeding the coercive field is applied to a specimen of multi-domain ceramic to produce a net remanent polarization. To explain this process in more detail, let us examine a few grains of the crystal as in Fig. 3.3. It has been initially polarized in the negative direction, and each domain is polarized more or less in the downward direction. Thus the originally square (viewed sideways) ceramic block has become elongated vertically. If an electric field in the positive direction is gradually applied, the block will contract initially, since the field opposes the polarized direction. As the electric field becomes stronger, some of the poles in the grains will begin to reverse direction. At a certain voltage, the block will no longer be able to contract any further. This electric field is called the coercive field, indicated by E_c.

If the field strength is further increased, the ceramic block will start to expand. When all the poles have been reversed, the block can expand no further; the field for this condition is indicated as E_{max} in the figure. If the electric field is then reduced, the strain will keep decreasing until the electric field reaches zero. In the final state shown, the poles in all the grains are reversed from the initial state, and the block has been polarized in the positive direction.

If an a.c. voltage with an amplitude smaller than the voltage supplied for poling is applied to a ceramic body having a remanent polarization, the strain, although showing some hysteresis, is approximately proportional to the applied voltage (see Fig. 3.4).

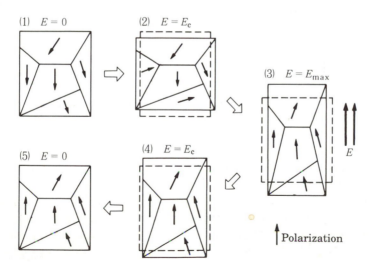

FIG. 3.3.   Diagrammatic representation of strains induced by pole reversals in a ferroelectric ceramic material.

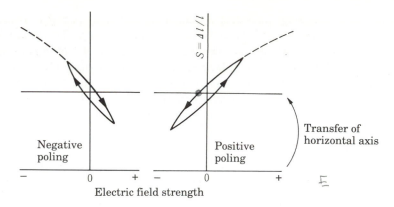

FIG. 3.4.   Strain vs. field strength for a polarized material.

3.2   Parameters for the piezoelectric effect

(1) *Piezoelectric strain constant*. As mentioned above, the strain and applied voltage are proportional in a polarized crystal. This relationship, if we ignore the hysteresis effect, can be expressed as

$$\text{For positively polarized state} \quad \frac{\Delta l}{l} = dE \qquad (3.2a)$$

$$\text{For negatively polarized state} \quad \frac{\Delta l}{l} = -dE. \qquad (3.2b)$$

The proportionality constant d is called the piezoelectric strain constant.

(2) *Poisson's ratio*. In the Langevin vibrator, strain in the direction of the electric field (i.e. the longitudinal effect) is utilized. This element, which will be discussed in full detail later, is used in the wedge-type motor (see Fig. 1.3) and also in a travelling-wave motor of the linear type (see Fig. 1.16b).

Piezoelectric strains are generally extremely small, with the strain constant d typically around 10^{-10} to $10^{-9}\,\mathrm{m\,V^{-1}}$. Taking $d = 10^{-9}\,\mathrm{m\,V^{-1}}$ for example, a $10\,000\,\mathrm{V}$ voltage applied on a ceramic specimen 1 cm thick creates an elongation of $0.1\,\mu\mathrm{m}$.

Consider instead the kind of behaviour seen when a flat block of Jello is squeezed together and expands sideways (see Fig. 3.5(a)). In a similar way, when an electric field is applied across the thickness of a ceramic body, it produces relatively large expansions at right angles to the field (the piezoelectric transverse effect). This is what drives the small ring-shaped motors.

Poisson's ratio is a parameter which indicates relative deformations in the longitudinal and transverse directions. Specifically, it is the ratio of

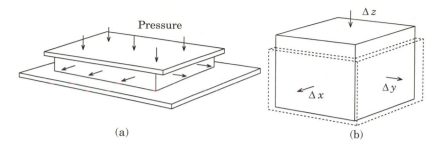

(a)   (b)

FIG. 3.5.   Illustration of Poisson ratio. (a) If a thin slab of Jello is pressed down, its area expands. (b) When a cube is pressed down in the z-direction and as a result a change Δz in length occurs, an expansion of $\Delta z/2$ will take place in both x-and y-directions, provided that the cube's volume is kept unchanged. In reality, the volume does not remain the same, and the ratio of the change in the x-direction to that in the z-direction is referred to as the Poisson ratio.

transverse elongation to longitudinal contraction when a pressure is applied to a solid at a constant voltage, or

$$\sigma^E = \frac{S_{31}}{S_{33}}. \tag{3.3}$$

The superscripts and subscripts in the terms are commonly used tensor expressions when describing piezoelectric phenomena. They have the following meanings:

(1)   a superscript denotes a non-varying parameter during state changes; thus σ^E is Poisson's ratio when the applied voltage is kept constant;

(2)   Subscripts indicate axis directions for cause and effect.

The numbers 1, 2 and 3 correspond to axes x, y and z respectively (Fig. 3.6). Thus a pressure in the z-direction (cause) creates a strain $\Delta z/z_0$ in the

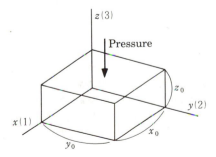

FIG. 3.6.   Coordinate axes and corresponding subscript numbers.

z-direction (effect), represented by S_{33}. Similarly, S_{31} is the strain in the x-direction caused by a pressure in the z-direction. Since normally $S_{32} = S_{31}$, S_{31} is often substituted for S_{32} when the latter is meant.

   A typical Poisson's ratio σ^E of piezoelectric materials is ~0.3 or a little higher (see Table 7.1, p. 179). So if a pressure causes a cube to contract $10\,\mu$m in the z-direction, the sides each expand by $3\,\mu$m. For a flat parallelepiped solid 10 mm wide by 10 mm long and 1 mm thick, the same contraction would cause the sides to expand by 10 times as much, or $30\,\mu$m.

(3) *Directionality of the piezoelectric strain constant.* If deformations are caused by an electric field, however, Poisson's ratio cannot be used for determining the relative deformations. In this case, the piezoelectric strain constants, which possess directional qualities as well, are used. The strain constant in the z-direction (for the longitudinal effect) is usually represented by d_{33}. That is,

$$\Delta z/z_0 = d_{33}E_z. \tag{3.4}$$

The elongation in the x-direction (for the transverse effect) is given by d_{31}. Thus,

$$\Delta x/x_0 (= \Delta y/y_0) = d_{31}E_z. \tag{3.5}$$

Let us consider an example, NEPEC-61, a material produced by Tokin Corporation. It has the following properties (see Table 7.1 (p. 179)):

$$\sigma^E = 0.31$$

$$d_{31} = -1.32 \times 10^{-10}\,\text{m V}^{-1}$$

$$d_{33} = 2.96 \times 10^{-10}\,\text{m V}^{-1}.$$

Therefore $d_{31}/d_{33} = -0.446$, which has a greater (absolute) magnitude than σ^E.

(4) *Voltage output coefficients.* The reverse piezoelectric effect is formulated as

$$E_z = -g_{33}T_z \tag{3.6}$$

$$E_z = -g_{31}T_{x,y}. \tag{3.7}$$

When measuring the induced voltage, from which the electric field is computed with the sample dimension, care must be taken not to generate a current. The proportionality constants g_{33} and g_{31} are called the voltage output constants. The g constants are closely related to the d constants, and have the simple relationships:

$$g_{33} = d_{33}/\varepsilon_{33}^T \tag{3.8}$$

$$g_{31} = d_{31}/\varepsilon_{31}^T \tag{3.9}$$

where ε is the permittivity (dielectric constant) of the ceramic. The units of g are $V\,m\,N^{-1}$.

3.3 Force factor and the piezoelectric equations

So far in our discussion, we have used terms such as strain, field strength, and stress. These terms, which express deformation per unit length, voltage per unit length, and force per unit area, respectively, are physical quantities which describe material behaviour from a general perspective. When discussing motors, however, quantities such as absolute displacement, voltage, force, and torque are used. The interrelationship of these engineering terms can be better understood if we introduce a force factor A, which is employed in electroacoustic transducer theory.

In describing the behaviour of systems that employ piezoelectric ceramics, the following equations are important:

$$-F = AV - Zv \tag{3.10}$$

$$I = Y_{\mathrm{d}}V + Av \tag{3.11}$$

where F is the force at the mechanical terminals, v the velocity at the mechanical terminals, V the voltage at the electrical terminals, I the current

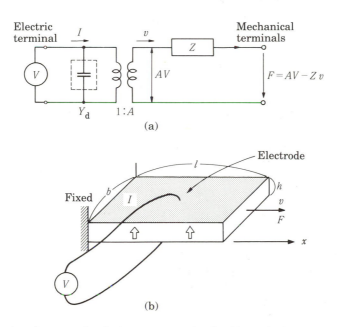

FIG. 3.7. An electromechanical transducer circuit: (a) equivalent circuit; (b) the transverse effect.

at the electrical terminals, Z the mechanical impedance of the ceramic material, and Y_d the blocking admittance, or the admittance of the capacitor proper as a dielectric material which is measured while blocking the deformation of the material.

To illustrate these equations, the transverse effect in a ceramic element and an equivalent circuit are shown in Fig. 3.7. In eqn (3.10), the first term on the right-hand side is the force generated within the ceramic body, which is equal to the applied voltage V multiplied by A. If we subtract the impedance drop, which is the product of the internal mechanical impedance Z and velocity v, from this force AV, the result is the force exerted by the motor externally. Thus the force factor A is the force created when a unit voltage is applied.

Let us consider eqn (3.11). $Y_d V$ is the current flowing through the capacitor and Av is the current which causes the ceramic body to deform. (A capacitor that is used as a regular dielectric is called a blocking admittance; this will be explained later.) Thus the force factor A is the current generated when unit velocity is imparted to the ceramic body. The units of A are $N\,V^{-1}$ or $C\,m^{-1}$, both being equivalent. In general, A depends on the properties and dimensions of the ceramic and how it is used. Here we use the model in Fig. 3.7(b) to derive an expression for A.

Thus the following relationships exist:

$$\text{Field strength} \quad E = V/h \tag{3.12}$$

$$\text{Stress} \quad T = F/bh \tag{3.13}$$

$$\text{Strain} \quad S = x/l \tag{3.14}$$

$$\text{Velocity} \quad v = dx/dt = j\omega x = j\omega\, Sl \tag{3.15}$$

$$\text{Current} \quad I = bl\, dD/dt = j\omega bl D \tag{3.16}$$

where h, b, and l are the thickness, width, and length respectively of the ceramic element, x is the displacement, and D is the flux density or electric displacement. j is $\sqrt{-1}$.

If we substitute these equations into eqns (3.10) and (3.11), we obtain

$$-Tbh = AEh - j\omega ZSl \tag{3.17}$$

$$j\omega lbD = Y_d Eh + jA\omega Sl. \tag{3.18}$$

So

$$T = j\omega Z \frac{l}{bh} S - \frac{A}{b} E \tag{3.19}$$

$$D = \frac{A}{b} S + \frac{Y_d}{j\omega} \frac{h}{bl} E. \tag{3.20}$$

In Section 3.2, we defined the piezoelectric constants d by eqns (3.4) and (3.5), and g by eqns (3.6) and (3.7). Besides these, many other constants are known. The following two well-known equations correspond to eqns (3.20) and (3.19) above:

$$\Delta D = eS + \varepsilon \Delta E \tag{3.21}$$

$$\Delta T = cS - e\Delta E. \tag{3.22}$$

These are known as the *e*-form piezoelectric equations and express the characteristics of piezoelectric ceramics. Comparing the two sets of equations, we obtain for the force factor A,

$$A = be. \tag{3.23}$$

We can express e in another form. Consider the first term on the right-hand side of eqn (3.22). The coefficient c shows the relation between stress and strain when the electric field is kept constant (i.e $\Delta E = 0$); this is the Young's modulus by definition. (Although we shall use E for the Young's modulus in Chapters 4, 5 and 7, Y_{11} is used for the x-directional Young's modulus of a ceramic in this chapter.)

On the other hand, d, the piezoelectric strain constant, expresses the relation between the field strength E and strain S while no stress is being applied (i.e. $T = 0$). Thus, from eqn (3.22), we obtain

$$d_{31} = \frac{S}{\Delta E} = \frac{e}{c} = \frac{e}{E_y}. \tag{3.24}$$

Hence

$$e = d_{31} Y_{11} \tag{3.25}$$

and so the force factor A is given by

$$A = bd_{31} Y_{11}. \tag{3.26}$$

It should be emphasized again that the force factor is generally determined not only by the dimensions and material properties of the piezoelectric ceramic element, but also by the properties of the metal plate with which it is combined. Equation (3.26) represents the force factor for a particular simple case. We can simplify the equivalent circuit of Fig. 3.7(a)

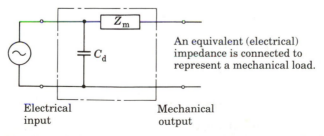

An equivalent (electrical) impedance is connected to represent a mechanical load.

Electrical input

Mechanical output

FIG. 3.8.   Simplified circuit for Fig. 3.7(a). Output terminals are considered to be short-circuited when no mechanical loads are connected.

by incorporating the transformer element into the internal impedance Z. The result, with a modified internal impedance Z_m, is shown in Fig. 3.8.

3.4 A.c. voltage application and resonance

We shall now examine the impedance Z_m in Fig. 3.8 to show that this circuit is a resonant circuit.

3.4.1 Equivalent capacitance

When a d.c. voltage is applied with no mechanical load, the piezoelectric element will reach equilibrium after undergoing an initial deformation. This is analogous to a spring under a constant force, which can be expressed by

$$F = Kx \tag{3.27}$$

where F is the applied force, x the displacement and K the spring constant.

The force generated in a piezoelectric element is AV, as indicated earlier. We also showed that A, the force factor, is equal to the current created when unit velocity is imparted. Or, integrating the terms with respect to time, it is the charge created by a unit displacement. Thus

$$q = Ax \tag{3.28}$$

Substituting these two relationships into eqn (3.27), we obtain

$$AV = K(q/A) \tag{3.29}$$

or

$$V = \frac{K}{A^2} q. \tag{3.30}$$

The equivalent capacitance is then given by

$$C = A^2/K. \tag{3.31}$$

The spring constant K in these equations cannot be determined solely by the piezoelectric ceramic's properties. In cases when the ceramic element is bonded to metal components, as in the dimorphic or monomorphic types, shown in Fig. 3.16 later, the metal properties must also be taken into consideration.

3.4.2 Equivalent inductance

Consider now the forced vibrations (i.e. elongations and contractions) of a piezoelectric ceramic element when an a.c. voltage is applied. If the

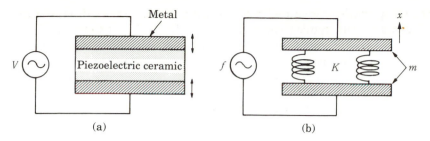

(a)   (b)

FIG. 3.9. a.c. voltage applied to a combined piezoeletric-ceramic–metal element and a mechanically equivalent model: (a) electrical–mechanical transformation; (b) mechanical model.

ceramic element is bonded to metal, then obviously the metal will also vibrate. Thus the vibrational magnitude and phase will be affected not only by the spring constant K, but also by the mass m of the ceramic-and-metal assembly.

A model depicting this behaviour is shown in Fig. 3.9, where (b) shows the forced vibration of the mass–spring system under a force f, generated by the a.c. voltage V. To construct an equivalent circuit, we represent the mass by an inductance L and place it in series with capacitor C, derived earlier. This is shown in Fig. 3.10. In this figure, the inductance L is given by

$$L = m/A^2. \tag{3.32}$$

This is briefly explained below. (The capacitance was given in eqn 3.31.)

In Fig. 3.9(b), the equation of motion is given by

$$m \frac{d^2x}{dt^2} + Kx = f(x). \tag{3.33}$$

If we substitute eqn (3.28) into this equation and express it in terms of q, then divide both sides by A, we obtain

$$\frac{m}{A^2} \frac{d^2q}{dt^2} + \frac{K}{A^2} q = \frac{f(x)}{A}. \tag{3.34}$$

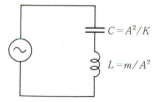

$C = A^2/K$

$L = m/A^2$

FIG. 3.10. Equivalent circuit for the model in Fig. 3.9.

Recalling the interpretation in Fig. 3.7, we see that the term on the right-hand side is the applied voltage.

On the other hand, we know that the equivalent capacitance C, equivalent inductance L, and voltage $V(t)$ are related by

$$L \frac{d^2q}{dt^2} + \frac{1}{C} q = V. \tag{3.35}$$

Since the terms containing the second derivative in eqns (3.34) and (3.35) must be equal, eqn (3.32) follows. Equation (3.31) can be also be checked by comparing the terms for q.

3.4.3 Resonance

The deformations created when a piezoelectric ceramic element is subjected to an applied voltage are quite small. An effective method of creating relatively large deformations with low voltages is to take advantage of resonance phenomena. For a CL series circuit, the characteristic angular frequency ω_0 is given by

$$\omega_0 = 1/\sqrt{CL} = \sqrt{K/m}. \tag{3.36}$$

The characteristic frequency ν_0 is given by

$$\nu_0 = \omega_0/2\pi. \tag{3.37}$$

When a spring–mass system, such as the one shown in Fig. 3.9(b) (a bell for instance), is struck, the vibration (or sound) generated has a certain frequency. This is the characteristic frequency ν_0. In free vibrations, the initial impact energy is transformed into the kinetic energy $\frac{1}{2} mv^2$ of the mass m and the potential energy $\frac{1}{2} Kx^2$ in the spring; these two forms of energy alternate with each other, fluctuating at twice the rate of the system's characteristic frequency ν_0. We shall describe this process below, using Figs 3.11 and 3.12.

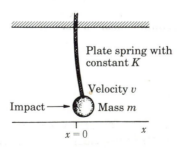

FIG. 3.11.   When kinetic energy is imparted to a mass connected to a spring, the system vibrates at the characteristic frequency.

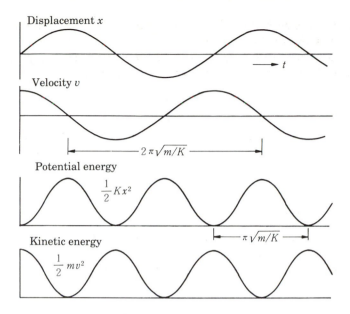

FIG. 3.12.   Free vibrations of a spring–mass system: fluctuations of displacement x, velocity v, potential energy, and kinetic energy.

After the impact imparts to the mass an initial velocity v, the mass decelerates as the plate spring is bent. The initial kinetic energy $\frac{1}{2} mv^2$ is gradually converted into the potential energy $\frac{1}{2} Kx^2$. When the velocity of the mass reaches zero, the kinetic energy has all been changed into potential energy (see Fig. 3.12). The movement is subsequently reversed.

The mass is given an acceleration in the opposite direction by the reaction force created in the bent plate spring. At the moment the spring returns to its initial position, the potential energy is zero, while the kinetic energy is at its maximum. The mass will continue its movement in the same direction owing to its inertia, all the time being slowed down by the plate spring's pull, until its velocity becomes zero once again, at which point its potential energy reaches its maximum level. After this, the spring will once again exert a force in the positive x-direction, trying to restore its shape, and the cycle is repeated.

If an impact is repeatedly given to the mass m at a rate equal to the natural frequency $(\frac{1}{2}/\pi)\sqrt{K/m}$ of the system, resonance is created and the amplitude increases.

Similarly, if a device which includes a piezoelectric ceramic element is excited at a frequency sufficiently close to its characteristic frequency, the vibrations are likely to reinforce each other and increase the amplitude. Under actual conditions, the amplifying effect is limited by viscous effects (caused by motion of the parts involved) or resistance to pole reversals within the crystal grains, so that the amplitude is kept below a certain level.

Resonance in an electric circuit occurs between a capacitor and an inductor (or reactor): electrical energy alternates between electrostatic energy $\frac{1}{2}Cq^2$ (stored in the capacitor) and electromagnetic energy $\frac{1}{2}Li^2$ (stored in the inductor). Such circuits able to store energy are called tank circuits: energy is stored in the 'tanks'.

On the other hand, the mechanical energy in piezoelectric ceramic systems, created via the electromechanical energy transformation, is stored as kinetic energy and the potential energy of strain (or strain energy). In the equivalent circuit these are represented by electromagnetic energy and the capacitor's electrostatic energy respectively. We shall resume the topics of equivalent circuits and resonance in Section 3.7. In the following section, we examine the vibrator and its mechanisms.

3.5   The bolt-tightened Langevin vibrator

The Langevin vibrator was developed in 1922 by the French scientist, Paul Langevin, to be used in submarine sonar prospecting. This vibrator makes use of the piezoelectric longitudinal effect and has a very high conversion efficiency.

As already mentioned, longitudinal deformations in a ceramic element are very small. The Langevin vibrator is constructed so as to amplify these minute deformations: two disc-type ceramic elements are sandwiched between three electrode plates, and then between aluminium cylinders which are bolted together (Fig. 3.13). The total length is made equal to

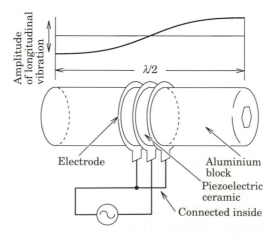

Fɪɢ. 3.13.   The bolt-tightened Langevin vibrator: piezoelectric ceramic discs are placed between two aluminium cylinders, which are then bolted together.

one-half the wavelength λ of the propagating longitudinal waves, maximizing the vibrational amplitudes at the ends of the aluminium cylinders (or metal blocks). Since sound waves travel through aluminium at a velocity of $5000 \, m \, s^{-1}$ at room temperature, for an excitation frequency of 50 kHz the wavelength can be determined to be $5 \times 10^6/(5 \times 10^4) = 100 \, mm$, so $λ/2 = 50 \, mm$.

Resonance in the vibrator amplifies the small deformations at the piezo-electric element (usually $<0.1 \, μm$) to create amplitudes as large as $5 \, μm$ at the vibrator's ends. This can be further increased four to ten times by means of a horn structure (see below).

The Langevin vibrator can also act as a generator; that is, if the vibrator's ends are vibrated by mechanical means, it will produce a voltage at the electrical terminals. The maximum voltage is obtained when the vibrator is vibrated at its resonant frequency. Figure 3.14 shows a diagram of a testing apparatus in which two Langevin vibrators are coupled: one acts as an oscillator, while the other acts as a generator (or absorber). A resistive load is used for the testing of vibrators.

In Chapter 5 we shall describe a travelling-wave motor in which two Langevin vibrators are used to generate travelling waves in a stator beam. One acts as the oscillator; the other acts as the absorber. This absorber is essentially the same as the generator in Fig. 3.14. The vibrations created by the oscillator at one end are absorbed at the other end to generate electric power, which is then dissipated as heat.

To amplify the vibrations further, a metal horn is attached as in Fig. 3.15. Although the horn should ideally have an exponential profile, the stepped type is often used since it is easier to machine. The length of the horn should be equal to $λ/2$. The amplification rate of the horn is given by the inverse ratio of the respective end surface areas.

The horn has the same effect as a transformer in a circuit: with a turns ratio of $n:1$, the current is amplified n times, while the voltage is lowered by $1/n$; in the horn, if the vibrational amplitudes are increased n times, the force is reduced by $1/n$.

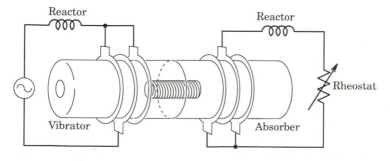

FIG. 3.14. Dummy load test using two Langevin vibrators.

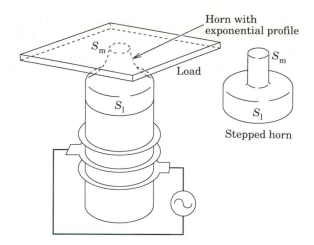

Fɪɢ. 3.15. Horn used for amplitude amplification; when S_1 and S_m are the areas of the larger and smaller cross-sections respectively, the multiplication factor n equals S_1/S_m.

3.6   Flexural waves created by the transverse effect

Although the Langevin vibrator has a high conversion efficiency as well as a high output, its relatively large size is unsuited for use in compact ultrasonic motors; a vibrator that utilizes a different principle must be used instead.

In Fig. 3.16 we show how a cantilevered elastic beam can be bent using piezoelectric ceramic elements owing to the transverse effect shown in Fig. 3.1(b). For the elastic beam, a metal such as aluminium or steel is used. The dimorphic type in (a) has piezoelectric elements bonded to both the upper and lower surfaces, whereas the monomorphic type in (b) has one element bonded to one side only. The polarization of the ceramic elements is indicated by arrows. Although the beam vibrates, no waves are generated.

Figure 3.17 shows how flexural waves can be created with a monomorphic structure. Note that the stator shown here is not a cantilevered beam; rather, it is considered as part of a free (or unrestrained) beam. The piezoelectric ceramic element is bonded on one side to the stator, which acts as one of the electrodes, and on the other side to the other electrode.

To create waves, it is important to place ceramic elements with opposite polarizations next to one another, so that one will expand while the other contracts when a voltage is applied (see (b) and (c) in Fig. 3.17). The metal stator will bend locally as the ceramic elements contract or expand, since they are bonded together; when the stator's lower surface expands, the upper surface will contract, and vice versa.

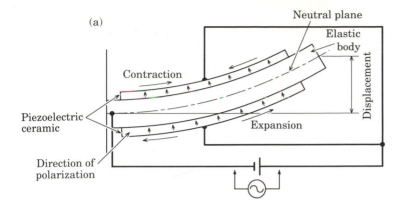

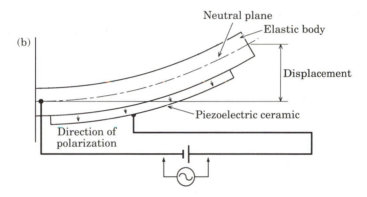

Fɪɢ. 3.16.   Principles of producing flexion on a lever, using the transverse effect:
(a) dimorphic type; (b) monomorphic type.

Flexural travelling waves are generated when two (or multiples of two)
ceramic elements A and B are positioned so as to have a 90° ($\pi/2$) phase
difference, and voltages proportional to $\sin \omega t$ and $\cos \omega t$ are applied to
phases A and B respectively (see Fig. 1.12, Chapter 1). Flexural waves will
be discussed in detail in Chapter 5.

3.7   The equivalent circuit

So far in this chapter, we have discussed some fundamental properties of
piezoelectric ceramics and shown how these ceramic elements are used in
vibrators. In this section we develop equivalent circuits to show how these
elements are applied in the ultrasonic motor.

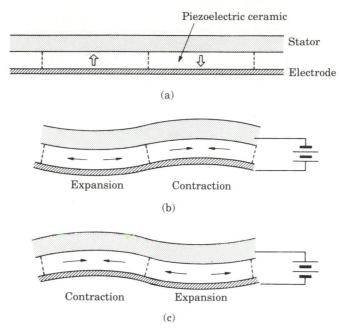

Fɪɢ. 3.17.   Principles of producing flexural waves in a beam in a monomorphic structure.

3.7.1   The basic equivalent circuit

By improving the equivalent circuit in Section 3.4, we obtain the circuit shown in Fig. 3.18. This circuit represents free vibrations of a stator with no loads, and includes two resistors which represent losses. The Langevin vibrator, the dimorphic and monomorphic plates shown in Fig. 3.16, and the single-phase behaviour of the stator beam of Fig. 3.17 can all be represented by this equivalent circuit. The elements that have been· added are given below.

C_d: This is capacitance due to the element's dielectric properties, called the 'blocking capacitance'. The term 'blocking' is explained in Fig. 3.19: the blocking capacitance is the capacitance measured when the ceramic element is fixed (or 'blocked') so that no vibrations can occur. In this condition there is no motion in the system. This is represented either as a break in the C_m–L_m branch (see figure) or an infinite resistance.

r_d: When an a.c. voltage is applied to a regular dielectric, the electric flux density (or electric displacement) D has a phase lag behind the electric field E. In other words, hysteresis is created between E and D and heat is generated. This is known as dielectric loss. Although this loss can be

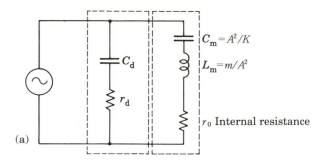

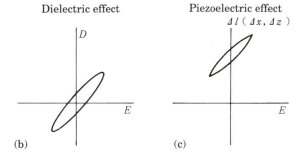

FIG. 3.18. Equivalent circuit with elements representing losses: (a) equivalent circuit; (b) loss represented by r_d (dielectric loss); (c) loss represented by r_0 (internal mechanical loss).

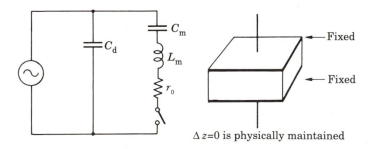

FIG. 3.19. Mechanically fixed (or 'blocked') ceramic element.

neglected at lower frequencies, at higher frequencies it cannot be ignored. The microwave oven is a cooking appliance which employs the heat generated by the dielectric loss of water.

The combination of C_d and r_d in the form $(r_d + 1/j\omega C_d)$ is called the blocking impedance. Since r_d can be neglected for the frequency range of the ultrasonic motor, we shall omit r_d in subsequent circuits.

r_0: As we saw earlier in Fig. 3.4, in the piezoelectric effect for a polarized element, the displacement (or strain) has a slight phase lag behind the electric field, creating a hysteresis loop which results in losses. Furthermore, various mechanical losses occur in components such as the vibrator's metal block, which is subjected to cyclic deformation. Losses are also created at the bonded surfaces of the piezoelectric ceramics, stator and electrodes. In the equivalent circuit, all the above losses are collectively represented by the element r_0, which is called the internal resistance. When r_0, C_m and L_m are combined in the form $(r_0 + j\omega L_m + 1/j\omega C_m)$, this is known as the motional impedance.

k: When a d.c. voltage is applied to a piezoelectric ceramic element (or Langevin vibrator, or ring-type motor without the rotor), electrical energy is stored in both C_d and C_m. Energy stored in C_m represents the strain energy of the ceramic element and metal components. The ratio of energy stored in mechanical form to the total energy provided by the source is equated with k^2. Its square root k is known as the electromechanical coupling coefficient:

$$k = \sqrt{\left(\frac{C_m}{C_d + C_m}\right)} \tag{3.38}$$

3.7.2 Resonant circuit parameters

Parameters employed in acoustic engineering are often used when discussing the piezoelectric ceramic element (by itself or combined with the stator) within the framework of equivalent circuits. We shall define only those parameters that are needed for our discussion, and later describe how to measure them. For a more extensive treatment of this subject, the reader should consult ref. 4.

(1) *Q-factor*. As stated earlier, to achieve high efficiencies the ultrasonic motor should be driven at (or close to) the frequency which will create resonance between C_m and L_m in the equivalent circuit of Fig. 3.20. The sharpness of the resonance is indicated by the Q-factor, which is given by

$$Q = \frac{\omega_0 m}{r_0} \tag{3.39}$$

where ω_0 is the resonant angular frequency, given by

$$\omega_0 = \sqrt{1/C_m L_m} = \sqrt{K/m}. \tag{3.40}$$

K is the spring constant for the combined system of piezoelectric ceramic element and metal, and m the combined (or equivalent) mass.

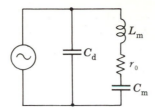

FIG. 3.20. The basic equivalent circuit.

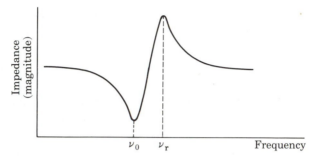

FIG. 3.21. Resonant frequency ν_0 and antiresonant frequency ν_r.

Greater Q values result from smaller r_0 values, or internal losses (i.e. losses at the stator only). The wedge-type motor has a Q-factor of ~ 500, while the ring-type travelling-wave motor has a Q-factor ~ 100.

(2) Resonance and antiresonance. At the resonant frequency, the impedance of the equivalent circuit in Fig. 3.20 has a minimum value. Conversely, there is a frequency at which the impedance has a maximum value. This is called the antiresonant frequency and is indicated by ν_r (see Fig. 3.21).

3.7.3 Measurement of parameters

Parameters for the equivalent circuit can be determined from ν_0, ν_r and their respective phase angles. These in turn can be obtained by measuring the magnitude and phase of the impedance while varying the excitation frequency. A schematic diagram for impedance magnitude and phase measurements is shown in Fig. 3.22. A large resistance R is placed between the voltage source and the ceramic element to maintain the a.c. voltage at a steady level. The impedance Z is then proportional to the terminal voltage of the ceramic element.

Using an oscilloscope, the impedance Z can be determined by measuring the voltage-to-current amplitude ratio, and the phase angle by the phase

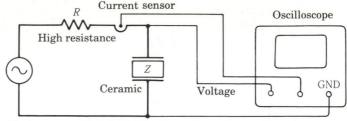

Variable-frequency source

FIG. 3.22.   Schematic diagram for measurement of impedance of piezoelectric ceramic element (alone or attached to stator).

difference between voltage and current. Figure 3.23 shows the results of a set of measurements. From ν_0 and ν_r, we can then determine the electromechanical coupling coefficient k by the following equation:

$$k \simeq \sqrt{\left[1 - \left(\frac{\nu_0}{\nu_r}\right)^2\right]}.$$ (3.41)

The parameters in the equivalent circuit can then be calculated from these measurements.

Since C_d is about two orders of magnitude larger than C_m, we can let $j\omega C_d$ approximate the circuit's admittance at lower frequencies (1 kHz for example). That is,

$$C_d = (\text{admittance at some low frequency } \nu)/2\pi\nu.$$ (3.42)

At the resonant frequency ν_0, the following relation is obtained:

$$C_m L_m = 1/(2\pi\nu_0)^2.$$ (3.43)

At the antiresonant frequency ν_r, on the other hand, the following equation is established, assuming a large Q-factor and a low r_0 value:

$$2\pi\nu_r C_d = \frac{1}{2\pi\nu_r L_m - \dfrac{1}{2\pi\nu_r C_m}}.$$ (3.44)

Solving the above two equations, we obtain

$$L_m = \frac{1}{(2\pi)^2} \frac{1}{C_d} \frac{1}{\nu_r^2 - \nu_0^2}$$ (3.45)

$$C_m = C_d \frac{\nu_r^2 - \nu_0^2}{\nu_0^2}.$$ (3.46)

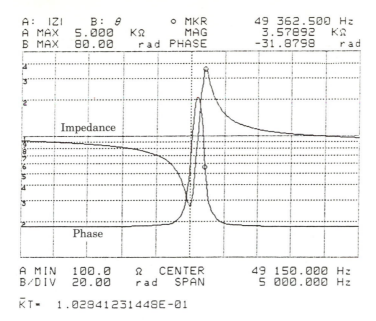

FIG. 3.23. Example of measurement of impedance magnitude and phase angle; points (○) indicate antiresonance.

Note that eqn (3.41) can be derived from eqns (3.38) and (3.46).

Finally, we can approximate r_0 by the impedance at resonance:

$$r_0 = \text{impedance at } \nu_0 \qquad (3.47)$$

3.8 Load connection

How can we express the load on the vibrator or piezoelectric element in the equivalent circuit? This is the question we shall address in this section. We begin by examining a Langevin vibrator driving a mechanism which includes friction, as shown in Fig. 3.24.

3.8.1 Incorporating load in the equivalent circuit

The mechanism shown in Fig. 3.24 is the same as in Fig. 2.19, Chapter 2. Using Z_L to indicate impedance, we can incorporate it into the vibrator's equivalent circuit as shown in Fig. 3.25. When discussing the quantity Z_L, force factor A becomes important.

The circuit in Fig. 3.25 consists of an electrical and a mechanical part. Note that the drive source supplies voltage instead of force. So the mech-

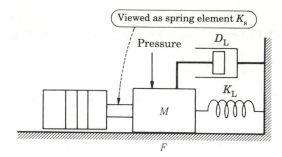

Fɪɢ. 3.24.   A Langevin vibrator drives a frictional load system.

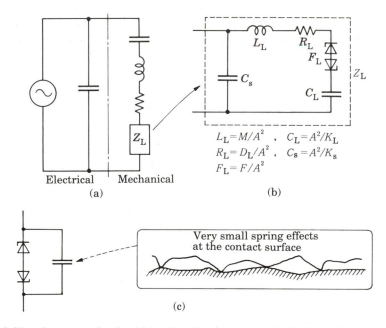

Fɪɢ. 3.25.   Incorporating load into the vibrator circuit: (a) basic equivalent circuit; (b) details of Z_L; (c) consideration of microscopic deformation.

anical circuit, which indicates force–velocity relationships, is represented by the equivalent electrical parameters of voltage and current; and mechanical impedance must be converted into electrical impedance using the force factor A. In Fig. 3.25, capacitor C_s represents the tip of the vibrator in contact with the load M acting as a spring. This effect however is small, so that a small capacitance would suffice.

Furthermore, for vibrations taking place at a microscopic level, small reversible (elastic) deformations can occur at forces less than the maximum static friction (refer back to Fig. 2.4). This effect is represented by a capacitor placed in parallel with the diodes (see Fig. 3.25c).

3.8.2 Load with rectifying effect

Consider the load in Fig. 3.26(a): the disc has grooves cut into its surface, and the vibrator tip makes contact with the disc's surface at nearly parallel angles, so that together they act as a ratchet mechanism. The tip pushes the disc in the counterclockwise direction when it catches a ratchet groove; in the reverse direction, the tip will merely slide over the disc. The equivalent circuit for this model is shown in (b).

The ratchet mechanism can be represented by a rectifier diode. When the vibrator tip extends (i.e. equivalent current flows in the positive direction), the equivalent diode becomes reverse-biased, the ratchet engages, and current (i.e. velocity) reaches the load. When the vibrator contracts, the diode

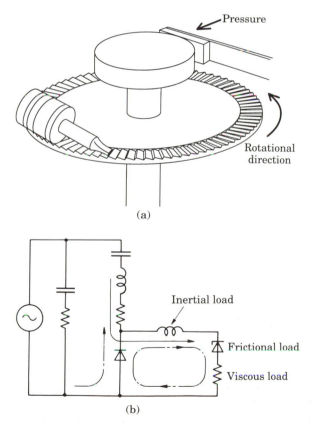

(a)

(b)

Fɪɢ. 3.26. (a) Rotor driven by ideal ratchet mechanism; (b) equivalent circuit.

becomes forward-biased, the vibrator and load continue their respective motions unrestrained, and current flows through the diode. The equivalent current flow is indicated by the chain lines.

The wedge-type motor, which we shall discuss in the next chapter, is similar to (but slightly more complex than) the model presented here.

3.8.3   *Impedance matching and the use of a horn*

In regular motors, e.g. a d.c. servomotor, it is important to match the motor's capacity with the load; that is, the motor and load should have about the same inertia, as in Fig. 3.27(a). Although the theory involved is fairly complex, we shall keep our discussion at a simple level. If the load's inertia is substantially lower than that of the motor, as in (b), much of the motor's power will be consumed in accelerating or decelerating its own inertia. Conversely, if the motor's inertia is lower than that of the load, as in (c), the power output is insufficient to rapidly accelerate (or decelerate) the load.

Consider next an electric circuit. In Fig. 3.28, a voltage source E which has an internal resistance r_i is connected to a load resistance R_L. The power consumption is given by

$$P_{IN} = \frac{E^2}{R_L + r_i} \tag{3.48}$$

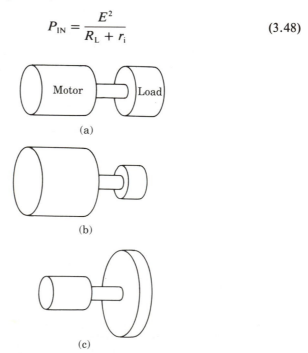

(a)

(b)

(c)

FIG. 3.27.   Matching motor–load inertias: (a) inertia is evenly matched; (b) motor has greater inertia; (c) load has greater inertia.

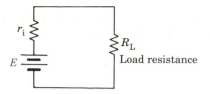

FIG. 3.28. Load resistance R_L connected to source with internal resistance r_i.

while the output P_{OUT} is given by

$$P_{OUT} = E^2 \frac{R_L}{(R_L + r_i)^2}.$$ (3.49)

Therefore the efficiency is determined by

$$\eta = \frac{R_L}{R_L + r_i}.$$ (3.50)

Two observations can be made:

(1) efficiency η is low if the load resistance R_L is smaller than r_i;

(2) if R_L increases, so does η; however, from eqn (3.49), this will reduce P_{OUT}, underutilizing the motor's capacity.

It is thus important to select a suitable load resistance R_L. (In determining R_L, other factors will need to be considered.) Returning to our discussion

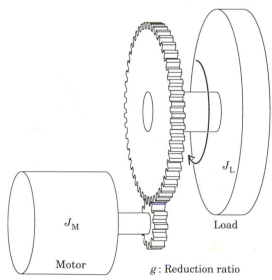

FIG. 3.29. Motor speed reduced to drive load.

on motors, motor and load inertias can be matched within a wider range
if a speed converter (e.g. gears or pulleys) is used. In Fig. 3.29, the reduction
ratio g (for speed conversion) that would minimize the motor's accelerating
torque is given by

$$g = \sqrt{(J_L/J_M)}. \tag{3.51}$$

This is one approach to matching motor and load inertias.

In the electric circuit, the transformer serves the role of gears. A transformer in an a.c. circuit consists of a core and coils, whereas in a d.c. circuit
it consists of a transistor (or some other switching element), a core, coils,
and a condenser. To simplify our discussion, we consider an a.c. circuit as
shown in Fig. 3.30. A load resistance R_L is connected to the source
through a transformer with a transformer ratio n, as in (a). This is
equivalent to a direct connection with load R_L/n^2. Therefore, if r_i and R_L
are known, the impedance can be matched by using a transformer with a
suitable transformer ratio.

We shall now see how impedance matching is done for the ultrasonic
motor. The models shown in Fig. 3.31 represent the circuits in Figs 3.28 and
3.30. A dashpot containing a highly viscous oil serves as the vibrator's load.
The internal impedance, which is represented by the resonant-system
elements L_m, C_m and a small internal resistance r_0, is connected to R in
series. Assume now that the vibrator is excited at slightly above the resonant
frequency, so that a steady a.c. current, whose amplitude I is determined
by the impedance $j(\omega L_m - 1/\omega C_m)$, flows in the circuit. Assume also that
R is small, so that the power output I^2R is much smaller than the source
(or vibrator) capacity. Thus we can improve the efficiency of this system
by matching impedance.

We can supply more power to the load by using a horn to amplify the
vibrator's motion, as in Fig. 3.31(b). Thus a transformer ratio of $1/n$ in the
equivalent circuit corresponds to an amplification of n: the voltage (or
force) is reduced by $1/n$, while the current (or velocity) is multiplied n times.

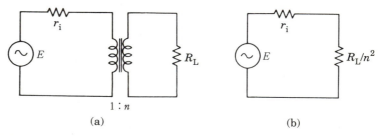

(a)   (b)

FIG. 3.30. Impedance matching in a.c. circuit is performed by a transformer
(transformer ratio n): (a) power supplied to load through ideal transformer;
(b) equivalent load resistance R_L/n^2.

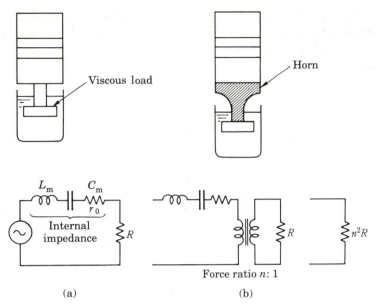

Fig. 3.31. Horn–transformer analogy: (a) load connected directly; (b) impedance modified by the horn.

In the circuit, the horn can be represented by transformer coils, or a resistance of n^2R can be directly connected to the source.

If we were to choose a very high value of n for our example, the impedance would become too large, the vibrator would slow down, and

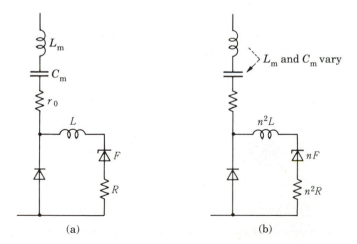

Fig. 3.32. The circuit of Fig. 3.26: (a) without horn; (b) with horn.

power transmission would be ineffective. If we were to include a horn in the model shown in Fig. 3.26 to increase the motor's speed, the result would be the circuit shown in Fig. 3.32.

3.8.4 Function of the comb-teeth in the ring-type motor

When the piezoelectric element is directly bonded to the stator (as in Fig. 3.16), it is not possible to attach a horn to match impedances. Instead, the comb-tooth structure performs this role. As shown in Fig. 1.7, Chapter 1, the comb-teeth are cut into the rotor side of the stator ring. The vibrational amplitude at the tip is larger than at the base of the comb-tooth. Because the comb-tooth structure provides the driving force, it is an important element in the travelling-wave motor. Without the comb-tooth structure, points on the stator's surface describe narrow elliptical trajectories (see Fig. 3.33), whereas the trajectories are nearly circular with the comb-tooth structure (see photograph in Fig. 3.34).

3.8.5 Impedance matching at the absorber

Above, we discussed impedance matching from the standpoint of motor design. In the linear ultrasonic motor, we need a mechanism at one end of

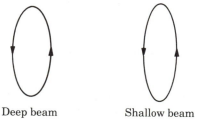

Deep beam   Shallow beam

FIG. 3.33.   Motion trajectory created by flexural travelling waves at beam surface.

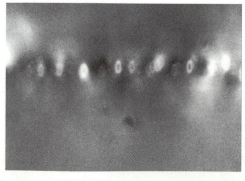

FIG. 3.34.   Motion trajectory at tip of comb-tooth.

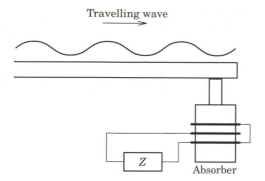

FIG. 3.35. Impedance Z must be matched in the linear motor so that the absorber will not reflect travelling waves (see Section 5.10).

the stator to absorb the wave energy and prevent standing waves from occurring. The Langevin vibrator can be used as a generator for this purpose. As shown in Fig. 3.35, the vibrator absorbs the vibration of the stator (or rail) to generate electricity, which is then consumed in the load circuit. The vibrator is called an absorber in this case. The electrical impedance of the absorber's load circuit must be adjusted properly so that waves will not be reflected. This will be discussed in greater length in Sections 5.4 and 5.10.

3.9 Output limits of the piezoelectric ceramic element

In the previous section, we discussed how loads can be matched to make effective use of the piezoelectric ceramic element. We next discuss output limits, which are closely associated with material characteristics.

3.9.1 Voltage and stress limits

As we showed in Section 3.1.4, a high voltage is applied to the piezoelectric ceramic to create residual polarization. This is analogous to the magnetization of a ferromagnet. To create the piezoelectric effect, an a.c. voltage of smaller amplitude is applied to the piezoelectric ceramic. How large a voltage amplitude can we apply? If the voltage exceeds a certain level, the voltage–strain relation will lose its linearity. Since the ceramic's polarization itself is not affected adversely by such low voltage levels, this does not create real problems in practice.

The upper and lower lines in Fig. 3.36(a) and (b) show the stress–strain relation at the positive and negative voltage limits respectively. There are also stress limits beyond which fatigue failure occurs in the ceramic material. The allowable compressive and tensile stresses are 5–6 MPa and

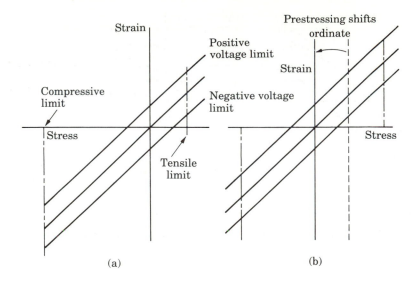

Fig. 3.36. Piezoelectric stress vs. strain characteristics: (a) without prestressing; (b) stress limit is increased with prestressing.

0.2–0.3 MPa respectively. These are indicated by chain lines in the graph; although (a) shows the compressive stress about three times the tensile stress, in reality the factor is 20–25.

The parallelograms in the figure show limits for voltage and force. Since an a.c. voltage has equal positive and negative amplitudes, the tensile limit becomes the critical factor. To increase the allowable stress range, therefore, the ceramic element is prestressed in compression (see Fig. 3.36b). In the Langevin vibrator, the bolt and nut provide prestressing. In the monomorphic element which utilizes the transverse effect (Fig. 3.17), the ceramic is bonded to metal which is thermally expanded; when the metal cools, prestressing occurs.

The theoretical maximum output is obtained when the stress or strain follows the parallelogram's perimeter in Fig. 3.36. With a sinusoidal a.c. voltage, however, we obtain instead an elliptical stress–strain path.

3.9.2   Actively matched state

When the stress–strain ellipse shows that the voltage and stress limits are both reached, this is called an actively matched state. This is rarely achieved in practice: either under- or over-mismatching occurs (see Fig. 3.37):

(1) under-mismatching: the stress limit is reached before the field strength limit;

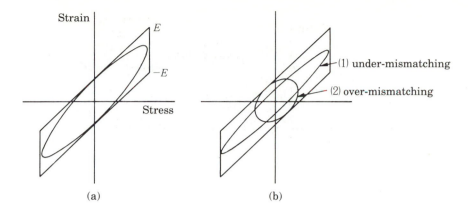

FIG. 3.37. Stress–strain ellipse for (a) actively matched and (b) mismatched conditions.

(2) over-mismatching: although the field strength limit is reached, the stress limit is not.

3.9.3 Fatigue limits

In Section 3.5, we described how the Langevin vibrator's metal block amplifies the ceramic element's vibrations. If the vibrational speed at the ends of the metal block exceeds certain limits, fatigue failure will occur.

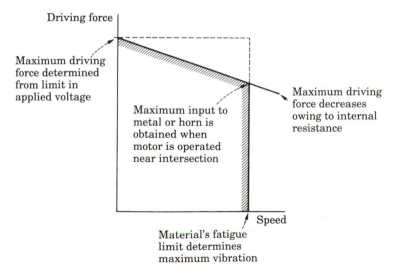

FIG. 3.38. Safe operating area.

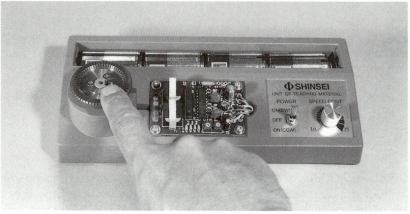

FIG. 3.39.   Educational ultrasonic motor kit. Top: the transparent plastic rotor will rotate when placed on the stator. Bottom: with the rotor removed, the driving force can be felt with the finger.

The amplifying effect of the horn is also limited by the metal's fatigue limit. Figure 3.38 illustrates the fatigue limit: the ordinate and abscissa represent the driving force and speed respectively. The driving force is determined by the applied voltage limit as well as in the internal resistance: for a set voltage, the driving force decreases with higher speeds owing to increasing internal resistance.

Ideally, the metal block or horn should be used so that the driving force and speed are both close to their respective limits.

3.9.4   Temperature limits

A piezoelectric ceramic loses its piezoelectric properties above a certain temperature. This is called the Curie point. The Curie point for NEPEC-61 (cf. Table 7.1) is 315°C; this is far above the ultrasonic motor's normal operating temperature and so presents no problems. Instead the practical temperature limit is often determined by weakening of the adhesive at higher temperatures.

In this chapter we have discussed the piezoelectric ceramic element as a drive source for the ultrasonic motor. In motor design, a single limiting factor usually becomes the bottleneck. When this limit is overcome by a design improvement, another limiting factor manifests itself. Thus the limits discussed above become important design parameters. The relation between the output and size of a motor will be discussed in Chapter 7.

References

1. Tanaka, T., Okazaki, K., and Ichinose, N. (1973). *Piezoelectric ceramics.* Gakkensha, Tokyo.
2. Uchino, K. (1986). *Piezoelectric and electrostrictive actuators.* Morikita, Tokyo.
3. Onoue, M., *et al.* (1982). *Fundamentals of solid vibration theory.* Ohmsha Ltd, Tokyo.
4. Ikeda, T. (1990). *Fundamentals of piezoelectricity.* Oxford University Press.

4. Theory and experiments on the wedge-type motor

In this chapter, we develop the theory of the wedge-type ultrasonic motor and present some experimental results. The high efficiency of the wedge-type motor was theoretically explained and experimentally demonstrated as early as 1982[1]. Yet to this day its significance has not been generally recognized. The wedge-type motor played an important role as a stepping stone to the development of the travelling-wave motor. Hence, if one is to understand the structure of the new motor and hope to improve on it, one must first gain a thorough grasp of the structure and theory of the original wedge-type motor.

4.1 Basic theory

A test model of the wedge-type motor was shown in Fig. 1.3 (Chapter 1). Its principles are illustrated in Fig. 4.1. There are three main components:

(1) the Langevin vibrator;

(2) a vibrator piece (or wedge tip) attached to one end of the vibrator;

(3) a rotor disc.

In the wedge-type motor, the vibrator piece P and rotor surface come into contact at a frequency of several tens of kilohertz. The vibrator piece is positioned at a slight angle θ from the normal to the rotor disc's surface, and its tip is slanted so that surface-to-surface contact is achieved with the rotor disc (see Fig. 4.1c).

We shall examine the motion of the tip of the vibrator piece P when vibrator V is being driven, taking the x-axis in the longitudinal direction and the y-axis normal to the x-axis. When the vibrator is oscillating, the base of the vibrator piece moves in the horizontal direction. At the tip, however, vertical movement is caused by the force F_t, which is the tangential component of the thrust force of piece produced owing to the angle of slant θ ($\sim 6°$) as shown in Fig. 4.1(c). Combining the synchronous harmonic motions in both directions, the tip will follow an elliptical path.

When the vibrator piece is pushed outwards by the vibrator, however, it will thrust the rotor and its tip will move from A to B along the rotor surface (see Fig. 4.2). From B to A, it will move unrestrained. It should be noted here that the material and dimensions of the vibrator piece are chosen so that its natural frequency matches the driving frequency of the

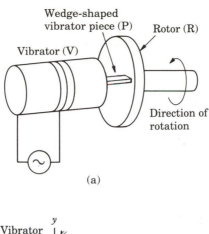

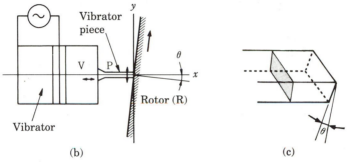

Fɪɢ. **4.1.** Basic structure of the wedge-type motor: (a) general view: (b) side view; and (c) vibrator piece tip.

vibrator. Moreover, we shall assume that the bending amplitude U_t at the tip is very small relative to the length of the vibrator piece. Using the x–y coordinate system, we shall examine the motion of the tip by dividing the path into two sections; B to A, where the vibration is unrestrained; and A to B, when the tip is in contact with the rotor surface.

4.1.1 Unrestrained (free) oscillation

The position (x, y) of the tip of the vibrator piece is given by the following equations:

$$x = U_0 \sin \omega t \qquad (4.1)$$

$$y = U_t \sin(\omega t + \alpha) \qquad (4.2)$$

where U_0 is the amplitude of the edge movement in the x-direction, U_t is the bending amplitude, ω is the angular frequency, and α is the phase

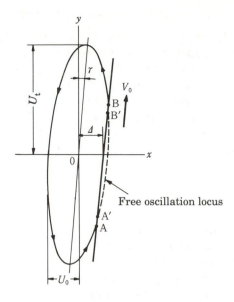

Fig. 4.2.   Trajectory of vibrator piece tip.

difference between x and y. By eliminating the variable t (time) from these equations, we obtain an equation for the elliptical trajectory:

$$\frac{x^2}{U_0^2} - \frac{2xy}{U_0 U_t} \cos \alpha + \frac{y^2}{U_t^2} = \sin^2\alpha. \tag{4.3}$$

The angle γ between the major axis of the ellipse and the y-axis is expected to equal θ and we can derive the following equation:

$$\tan 2\theta = \frac{2U_0 U_t}{U_0^2 - U_t^2} \cos \alpha. \tag{4.4}$$

As the speed increases, U_t becomes much greater than U_0, α approaches 0, and we have the relationship between U_0, U_t, and θ, expressed as follows:

$$\tan \theta = U_0/U_t. \tag{4.5}$$

4.1.2   Motion during contact

During contact, the tip moves along the rotor surface, the equation for which is given as a function of x by

$$y = \frac{1}{\tan \theta} (x - \Delta). \tag{4.6}$$

If the tip contacts the rotor at $\omega t = \beta$, Δ is given by

$$\Delta = U_0 \sin\beta - U_t \sin(\alpha + \beta)\tan\theta \qquad (4.7)$$

Next, we consider the y component U_y of the velocity of the vibrator piece tip. The tip comes in contact with the rotor surface at point A. At that point, it has a velocity component of $U_y = U_t \omega \sin\alpha$, and is moving more slowly than the rotor. The tip is then accelerated to a speed of $V_0/\cos\theta$ at point A', where V_0 is the rotor velocity at the contact surface. After this, the tip engages with the rotor surface by friction and moves along at the same speed as the rotor from A' to B'. Beyond point B', the tip gradually losses its frictional grip and decelerates until it separates from the surface at point B.

4.1.3   Forces at the vibrator piece tip

In this section we discuss the dynamics of forces at the vibrator piece tip. If allowed to vibrate under unrestrained conditions, the tip P will move along a path indicated by the broken line in Fig. 4.3. However, being restrained in the x-direction by the rotor surface, the vibrator piece undergoes a deformation δ in its longitudinal direction. If we let K_0 be the spring constant of the vibrator piece in the longitudinal direction, a force given by

$$F_r = K_0 \delta \qquad (4.8)$$

is experienced by the vibrator piece, which is counterbalanced by the force generated by the vibrator minus the force needed to accelerate it. This force is applied to the rotor surface at an angle θ from its normal. If we let F_n

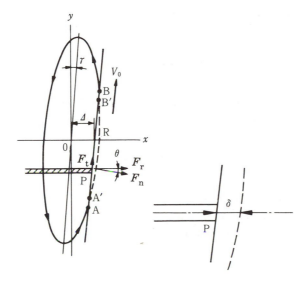

FIG. 4.3.   Force and trajectory of vibrator piece tip.

and F_t be the force components in the normal and tangential directions respectively, then

$$F_n = F_r \cos \theta = K_0 \cos \theta \cdot \delta \qquad (4.9a)$$

$$F_t = F_r \sin \theta = K_0 \sin \theta \cdot \delta \qquad (4.9b)$$

and F_t becomes the driving force of the rotor.

4.1.4   Equivalent-circuit analysis

In this section we use the equivalent-circuit theory introduced in Chapter 2 to examine the mechanism of the forces and motions that are involved. The equivalent circuit shown in Fig. 4.4(a) illustrates the role of each motor component. Switch S_1 represents contact and separation between the vibrator piece and the rotor's surface. The deformation δ of the vibrator piece is represented by the charge stored in capacitor C_1. The force F_r in eqn (4.8) is represented by the terminal voltage on C_1.

The equivalent circuit shows that the force generated by the piezoelectric ceramic, less the force necessary for accelerating the mass (represented by L_m) of the vibrator–vibrator-piece assembly, is in equilibrium with F_r (represented by the voltage on C_1). The force transformation takes place when F_r is transformed to F_t, which acts to bend the vibrator piece and drive the rotor. The transformation ratio is $1 : \sin \theta$.

After C_1 discharges (i.e. vibrator piece separates from rotor), it cannot become charged in reverse polarity (i.e. δ cannot be a negative value). For this purpose, diode D_1 is placed in parallel with C_1. The time taken from separation to contact again depends on Δ(see Section 4.1.2), which is not discussed in this book but is thought to be related to such factors as the torque on the load and the applied pressure.

L_v and C_v are elements representing the flexural vibrations of the vibrator piece, which are chosen so that the resonance frequency matches the vibrator's excitation frequency. L_R, R_R, D_R and C_R are all related to the rotor. By eliminating the transformer using the principle shown in Fig. 3.31 with parameter conversions in the secondary circuit, we obtain the circuit in Fig. 4.4(b).

If we assume that the rotor is stationary (i.e. no current flows in the circuit), when the vibrator piece and rotor are in contact and the piezoelectric ceramic is exerting a constant force, then the inductors, resistors and diodes can be removed to obtain Fig. 4.4(c). Here it is also assumed that no static friction exists. F_n can then be measured at the load (or torque) meter as $F_r \sin \theta$ due to the force transformation. These assumptions are true only if there are no losses. When the motor is running, the measured force will be less than $F_r \sin \theta$ because of the following factors:

• friction exerted on the bearing,

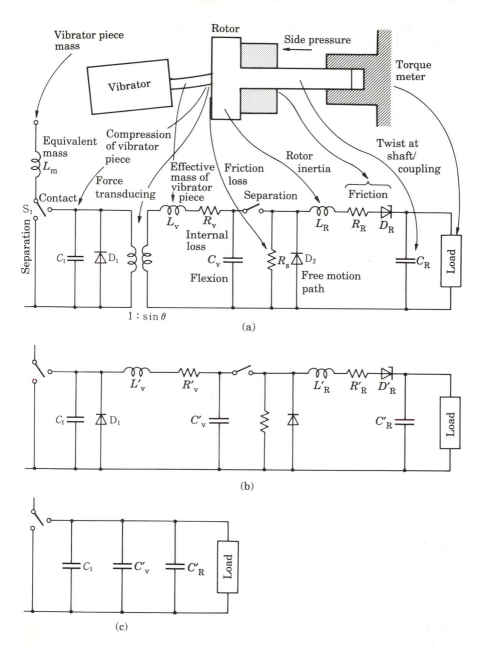

Fig. 4.4. Equivalent circuit for analysing transmission of forces and motions: (a) constructing an equivalent circuit; (b) when the transformer representing force conversion is simplified; (c) static model representing motor stopped with tip and rotor in contact.

- internal loss in the shaft and vibrator piece
- losses due to sliding at the contact surface.

Resistance R_s (or R_s'), which represents losses due to sliding, is a non-linear quantity varying with δ (the charge stored in C_1) and the rotor's speed (current in L_R or L_R'). A finite R_s indicates that sliding occurs, whereas an infinite R_s means that no sliding takes place.

We can assume that sliding occurs when F_t exceeds the frictional force to become

$$F_t = \mu F_n. \tag{4.10}$$

We can expect sliding, and the losses that it causes to be small owing to the small angle θ. Thus if losses due to viscosity and friction at other parts (i.e. resistors or diodes in the circuit) are minimized, the wedge-type motor should achieve high efficiencies.

In Fig. 4.4(b), if the load torque, which is represented by the load voltage, is decreased, the equivalent current increases. This means that speed increases as torque decreases.

When the side pressure on the rotor is increased, this will result in a higher F_n and higher frictional forces, and increase the sliding period (when eqn (4.10) applies). Hence the efficiency is expected to be improved. However, this will also result in higher R_R and D_R values, and overall losses will increase after side pressure is increased above a certain critical value.

In this section, the equivalent circuit has been presented to familiarize the reader with the basic elements of the ultrasonic motor. In the following sections we discuss these in more detail and also present the results of some laboratory tests. Interested readers may find it useful, however, to perform a more detailed equivalent-circuit analysis and compare the results with those from the tests.

4.2   The test model

A test model of the wedge-type motor was constructed to determine some of the motor's characteristics (see Fig. 4.5). A 28 kHz PZT Langevin vibrator was used as the vibratory source, to which a stepped horn with a transformation ratio of 1:2 was attached. The vibrator piece was then attached to the end of the horn with bolts, and a microscope was set up to observe the vibrator piece tip. The variable side pressure was applied with a spring, while the rotor was supported by a bearing and connected to a torque meter by a coupling device. The torque meter carried a tachometer. Painstaking care was given to the assembly of the test model, some of which is highlighted in the following two sections.

(a)

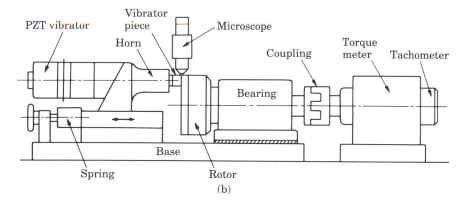

(b)

FIG. 4.5. (a) Test model of the wedge-type motor; (b) its component arrangement.

4.2.1 Rotor material

The rotor surface must have a high coefficient of friction to establish firm
contact with the vibrator piece, so that power can be transmitted efficiently.
At the same time, the rotor material should be highly resistant to abrasion.
In this model, commonly available materials, such as gunmetal and porous
bearing metal, were used. The coefficient of friction μ varies greatly with

surface lubrication conditions. The tests were performed with the surface coated with a fine film of lubricant to achieve a boundary lubrication state (i.e. a layer of lubricant one molecule thick).

Furthermore, the gap between the vibrator piece and the rotor surface must be kept at a distance of a few micrometres. Because of thermal expansion and stress deformation it is very difficult to maintain such small distances accurately. However, we are able to accomplish this under certain conditions by applying a constant pressure on the rotor during vibration.

4.2.2 Vibrator piece (wedge)

The vibrator piece is the most critical component of the wedge-type motor. To achieve high output and efficiency, the vibrator's longitudinal motion must be converted into a nearly perpendicular motion. At the same time, sliding between the vibrator piece and the rotor surface should be kept to the minimum.

The vibrator piece experiences large bending stresses at the base, while its tip undergoes repeated frictional contact. The material should therefore resist fatigue and stand up well to abrasion. In this model, we used carbon tool steel (high-speed steel) which was hardened and tempered to achieve a Rockwell hardness of >60. This was formed into a flat plate with a constant width h. If we consider the vibrator piece as a cantilevered beam, its fundamental natural frequency ν_1 is given by

$$\nu_1 = \frac{1}{2\pi}\left(\frac{0.6\pi}{l}\right)^2 \sqrt{\frac{EI}{\rho bh}} \tag{4.11}$$

where E is the Young's modulus, ρ the density, l the length, h the thickness, b the width, and I the second moment of the area. In this case we have

$$I = \frac{bh^3}{12}. \tag{4.12}$$

Equation (4.11) is derived in Section 5.9.2, and I is defined by eqn (5.63).

Substituting eqn (4.12) into eqn (4.11), we obtain

$$\nu_1 = 0.163\frac{h}{l^2}\sqrt{\frac{E}{\rho}}. \tag{4.13}$$

If we substitute the values for the Young's modulus for carbon tool steel, $E = 22 \times 10^{10}\,\mathrm{N\,m^{-2}}$, and its density $\rho = 7.8 \times 10^3\,\mathrm{kg\,m^{-3}}$, then

$$\nu_1 = 8.66\frac{h(\mathrm{mm})}{l^2(\mathrm{mm^2})} \times 10^5\,\mathrm{Hz}. \tag{4.14}$$

The values of h and l are then selected so that the frequency given by this equation approximately equals 28 kHz, the natural frequency of the

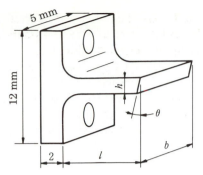

Fɪɢ. 4.6.   Vibrator piece.

Langevin oscillator. In this model, the vibrator piece was shaped as shown in Fig. 4.6, and 0.25, 0.5 and 1.0 mm were selected for the thickness h. From these values, using eqn (4.14) and setting $\nu_1 = 28$ kHz, l can be calculated to be 2.78, 3.93, and 5.56 mm respectively. The spring constants K_0 and K_1 in the longitudinal and bending directions respectively, are given by

$$K_0 = \frac{Ebh}{l} \qquad (4.15)$$

$$K_1 = \frac{Ebh^3}{4l^3} . \qquad (4.16)$$

4.3   Tests

Several tests were performed, including motion analysis of the vibrator piece and investigations of torque characteristics. In this section we present the results and analyse them.

4.3.1   Vibrator piece trajectory

The photograph in Fig. 4.7 shows the tip of the vibrator piece in contact with the rotor. The photograph in Fig. 4.8 was taken through a microscope to check the trajectory of the vibrator piece tip. The rotor is shown on the right, while the vibrator piece is slightly tilted in the counterclockwise direction. The numerous elliptical trajectories appearing on the left are those of imperfections or tiny objects present on the surface of the vibrator piece. Motion in the major axis direction was caused by the bending vibrations of the vibrator piece, while that in the minor axis direction was caused by the vibrator's longitudinal vibrations. The ellipses were created

FIG. 4.7.   Vibrator piece touching the rotor disc.

by counterclockwise movement at a rate of 28 000 rotations per second and
a maximum speed of $3.6\,\mathrm{m\,s^{-1}}$.

Figure 4.9 shows the trajectories of the vibrator piece tip when the
rotor speed was varied by changing the load while the amplitude was
kept constant. If we compare U_t, the width of the trajectory's major axis,
and V_0, the rotor velocity, we see that an approximately linear relation
exists (see Fig. 4.10). In the graph, $v_{max}\,(=U_t\omega)$, maximum velocity for
unrestrained vibration, is shown for comparison. We can see that the rotor's
velocity V_0 is ~85% of v_{max}, indicating that sliding is relatively slight,

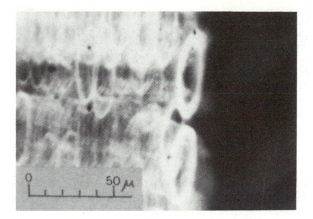

Fig. 4.8. Trajectory of vibrator piece at a vibrator frequency of 28 kHz and rotor velocity at contact point of 3.6 m s^{-1}. The trajectories of surface imperfections are also shown. In this experiment, the perpendicularity of the rotor surface to the shaft was so good that no surface fluctuations were observed.

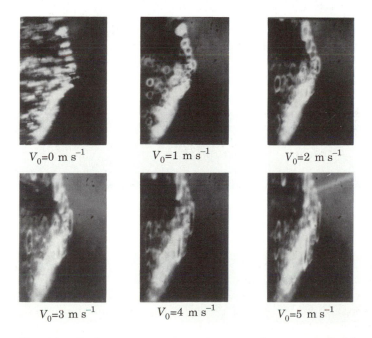

$V_0=0$ m s^{-1} $V_0=1$ m s^{-1} $V_0=2$ m s^{-1}

$V_0=3$ m s^{-1} $V_0=4$ m s^{-1} $V_0=5$ m s^{-1}

Fig. 4.9. Trajectory of vibrator piece at various rotor velocities.

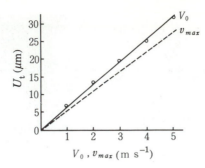

Fɪɢ. 4.10. Dependence of v_{max} $(= U_t\omega)$, the maximum velocity for unrestrained vibration, and V_0, the rotor velocity at the contact point, on U_t, the vibrator piece amplitude.

regardless of the speed. Thus we can conclude that losses caused by sliding at the contact surface are relatively small.

4.3.2 Measurement of contact period

Measurements were taken of the period during which the vibrator piece tip and the rotor were in contact. The results are shown in Fig. 4.11. The sine waves represent the vibrator's longitudinal displacements; the crests show the vibrator's position when closest to the rotor. The rectangular wave

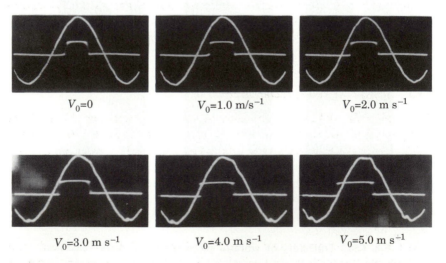

Fɪɢ. 4.11. Measurements of the contact period between vibrator piece and rotor disc: the sine waves represent the longitudinal displacement of the vibrator piece; the rectangular waves indicate contact and separation.

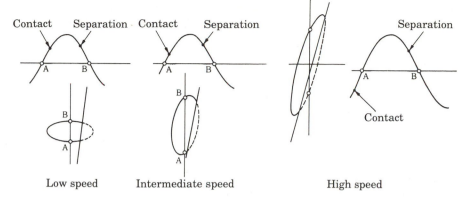

Fɪɢ. **4.12.** Mechanism to explain earlier contact/separation at higher velocities.

shows the level of the contact resistance between the vibrator piece and rotor: the higher level indicates contact, while the remainder indicates separation.

At lower velocities, we can see that contact and separation occur when the vibrator is closest to the rotor. At higher velocities ($V_0 = 3 - 5\,\mathrm{m\,s^{-1}}$), however, the moment of contact has shifted to the left (i.e. become earlier). Although the moment of separation has also shifted slightly earlier, the contact period has increased as a whole. This can be explained as follows (see Fig. 4.12). At higher velocities, the ellipse's major axis becomes closer to parallel with the rotor's surface. The positive phase of the vibrator's displacement is from point A to point B. The point of contact approaches A as the rotor's velocity is increased, and if the velocity is high enough, contact is made before point A.

4.3.3 *Speed–torque characteristics*

The graph in Fig. 4.13 shows the relation between velocity and driving force under varying conditions. A vibrator piece with thickness $h = 0.5\,\mathrm{mm}$ and width $b = 5\,\mathrm{mm}$ was used, and the side force on the rotor was varied between 13 and 27 N as shown. The driving force (mean value of F_t) and velocity V_0 are plotted on the abscissa and ordinate respectively. The angle of tilt was $\theta = 6°$ and the vibrational amplitude was $U_0 = 5\,\mu\mathrm{m}$. Measurements at low driving force and high velocity were not possible owing to the friction of the bearing.

As noted in the equivalent-circuit analysis in Fig. 4.4, the rotor's driving force should be equal to the applied force multiplied by $\sin 6° = 0.1045$ at the moment of start-up (i.e. no current flowing in L_R), if there are no losses. However, because of actual losses under running conditions at a high driving frequency of 28 kHz, the measured driving force will be lower than this. We can define the conversion rate as the measured result divided by the theoretical no-loss value. As shown in Fig. 4.14, the conversion rate is close to 100% when the applied pressure is low, but drops as the pressure is increased.

4.3.4   Output and efficiency

The graph in Fig. 4.15 shows the output power P_{OUT} derived from results shown in Fig. 4.13, and the efficiency η (involved in converting ultrasonic vibrations into a simple rotation), obtained by comparing P_{OUT} with the electrical input power (the ultrasonic output minus vibrator losses). From

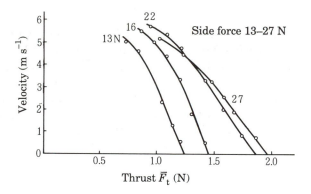

FIG. 4.13.   Relation between mean driving force (measured) and velocity.

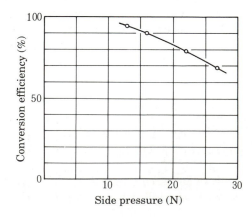

FIG. 4.14.   Relation between applied pressure and driving force conversion rate.

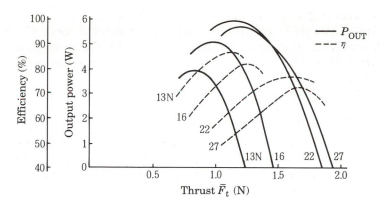

FIG. 4.15.   Output P_{OUT} and efficiency η measurements.

this graph, we can see that maximum efficiencies are obtained at lower speeds and higher driving forces. A maximum efficiency of 87% is obtained with a side force of 13 N. In Section 8.6.2 we shall present a theoretical proof showing that the travelling-wave motor also achieves higher efficiencies when driven at lower speeds. This is a major feature of the ultrasonic motor. (As a general rule, electromagnetic motors exhibit higher efficiencies at higher operating speeds.) To use this feature to our advantage, however, it is necessary to minimize the losses due to sliding at the contact surface.

4.4   The prototype model

Based on results obtained from the test model, we designed and built a prototype motor (see Fig. 4.16). A PZT Langevin vibrator of a 30 mm diameter is fitted inside a casing with an outer diameter of 40 mm.

One end of the vibrator is connected to a cylindrical horn, to which 36 vibrator pieces, each having a thickness of 0.1 mm, are attached (see Fig. 4.16). The rotor is in contact with the edges of the vibrator pieces,

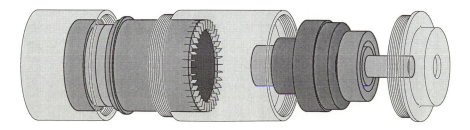

FIG. 4.16.   Schematic diagram showing how 36-piece vibrator units are assembled.

FIG. 4.17.   Wedge-type ultrasonic motor with clear plastics case.

and is supported by a central shaft. An automatic pressure device, which applies pressure proportional to the torque, is positioned between the rotor and shaft (refer back to Fig. 1.3).

The photograph in Fig. 4.17 shows a prototype model with a casing of clear plastics, making it easy to observe the inner mechanisms. A graph of its characteristics was presented in Fig. 1.6. The maximum efficiency of 60% is much lower than that of the test model shown in Fig. 4.5. This is because it is not possible to ensure optimum conditions for all 36 vibrator pieces at the same time.

Reference

1. Sashida, T. (1982). A prototype ultrasonic motor — principles and experimental investigations. *Oyobutsuri* (*Applied Physics*), **51**, 713–20.

5. Theory of the ultrasonic-wave motor

The ultrasonic wave motor was viewed as a major innovation compared with the wedge-type motor when it was announced, and it generated considerable interest in ultrasonic motors in general. In this chapter, the theory of flexural waves is first presented, then the mechanisms involved in generating thrust or torque from flexural waves are explained. The phenomenon of energy losses when electric power is converted to mechanical work is also discussed.

5.1 Basic wave equation

In Chapter 1, the concepts of standing waves and travelling waves were introduced, using waves generated in a stretched string. In the ultrasonic-wave motor, the waves are created by vibrations in a beam rather than a string. Before discussing waves in a beam, however, it will be useful first to gain an understanding of waves carried in a string. Consider a uniform string stretched along the x-axis (see Fig. 5.1a) which is made to vibrate in the y-direction (b). The behaviour of each point along the string can be determined by setting up and solving the appropriate equations of motion. Forces acting on a string segment are shown in Fig. 5.2.

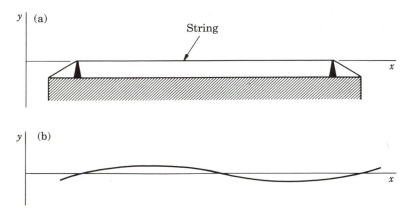

Fig. 5.1. Analysing vibrations in a string: (a) string stretched along the x-axis; (b) when a force is applied, each point vibrates slightly in the y-direction (displacements are exaggerated).

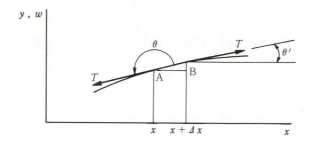

Fig. 5.2.   Forces acting on a small string segment Δx.

Let $w(x, t)$ be the displacement of point A at time t, and assume that the tension T is uniform along the entire length of the string. Then the tension acting on a small segment AB of the string has a w-component given by

$$T \sin \theta' + T \sin \theta. \tag{5.1}$$

When θ is approximately π ($= 180°$) and θ' is small, however, the following approximations may be made:

$$\sin \theta \simeq \tan \theta = -\frac{\partial w}{\partial x} \tag{5.2}$$

$$\sin \theta' \simeq \tan \theta' = \frac{\partial w}{\partial x} + \frac{\partial^2 w}{\partial x^2} \Delta x. \tag{5.3}$$

Therefore, we obtain

$$T \sin \theta' + T \sin \theta = T\frac{\partial^2 w}{\partial x^2} \Delta x. \tag{5.4}$$

If the mass per unit length (i.e. the linear density) of the string is ρ, then section AB has a mass of $\rho\Delta x$. Since the acceleration is $\partial^2 w/\partial t^2$, the equation of motion of this segment is given by

$$\rho\Delta x \frac{\partial^2 w}{\partial t^2} = T\frac{\partial^2 w}{\partial x^2} \Delta x. \tag{5.5}$$

By eliminating Δx from both sides, we obtain the following basic equation:

$$\frac{\partial^2 w}{\partial t^2} = \frac{T}{\rho} \frac{\partial^2 w}{\partial x^2}. \tag{5.6}$$

This is the equation of vibration (or wave equation) for the string. The motion of the string can be determined if this equation is solved and the function $w(x, t)$ is identified.

Equation 5.6 can be rewritten as

$$\frac{\partial^2 w}{\partial t^2} = v^2 \frac{\partial^2 w}{\partial x^2} \tag{5.7}$$

where

$$v^2 = T/\rho. \tag{5.8}$$

5.2 String vibrations and standing waves

When both ends of a string are fixed,

$$w = 0 \quad \text{at } x = 0 \tag{5.9a}$$

$$w = 0 \quad \text{at } x = l. \tag{5.9b}$$

Assume that the string has a frequency of $\omega/2\pi$ and that the solution for eqn (5.7) be given in the following form:

$$w = (C \cos \omega t + D \sin \omega t) q(x). \tag{5.10}$$

Substituting this into eqn (5.7), we obtain

$$-\omega^2 q = v^2 \frac{d^2 q}{dx^2} \tag{5.11}$$

or

$$\frac{d^2 q}{dx^2} + k^2 q = 0 \tag{5.12}$$

where

$$k^2 = \omega^2/v^2. \tag{5.13}$$

For the boundary conditions, we have

$$q = 0 \quad \text{at } x = 0 \tag{5.14a}$$

$$q = 0 \quad \text{at } x = l. \tag{5.14b}$$

The general solution for eqn (5.12) is

$$q = \alpha \sin kx + \beta \cos kx \tag{5.15}$$

where α and β are undetermined coefficients. Since the second term does not satisfy condition (5.14a), $\beta = 0$ and we obtain

$$q = \alpha \sin kx \tag{5.16}$$

From the condition at $x = l$,

$$kl = m\pi, k = \frac{m\pi}{l} \quad (m = 1, 2, \ldots). \tag{5.17}$$

Thus k has discrete values and is called the characteristic value of eqn (5.12). The solution of eqn (5.6) thus becomes

$$w = w_m = (A_m \cos \omega_m t + B_m \sin \omega_m t) \sin\left(\frac{m\pi}{l}x\right) \qquad (5.18)$$

where

$$\omega_m = v \frac{m\pi}{l}. \qquad (5.19)$$

The frequency is given by

$$\nu_m = \frac{\omega_m}{2\pi} = \frac{m}{2l} v = \frac{m}{2l} \sqrt{\frac{T}{\rho}}. \qquad (5.20)$$

Equation (5.18) can be expressed in the following form:

$$w_m = C_m \sin\left(\frac{m\pi}{l}x\right) \cdot \sin\left(\omega_m t + \gamma_m\right) \qquad (5.21)$$

where

$$C_m \sin \gamma_m = A_m \qquad (5.22a)$$

$$C_m \cos \gamma_m = B_m. \qquad (5.22b)$$

From eqn (5.21), we see that each point on the string is oscillating with an angular frequency of ω_m, and that its amplitude is a function of its position x. This is shown in Fig. 5.3.

Looking at the first term in eqn (5.21), we see that the amplitude becomes zero at

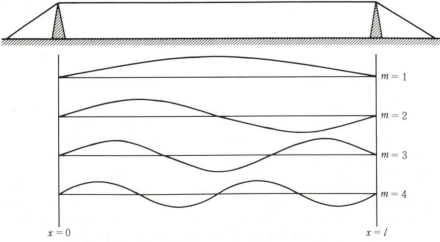

FIG. 5.3.   Standing waves at maximum amplitude.

$$x = \frac{n}{m} l \tag{5.23a}$$

which is called the node. The maximum amplitude is obtained at

$$x = \frac{\left(n - \frac{1}{2}\right) l}{m} \qquad (n = 1, 2, \ldots, m) \tag{5.23b}$$

which is called the loop (or antinode). This type of oscillation is called a standing wave.

5.3 Travelling waves

We now consider travelling waves by studying the general solution of eqn (5.7), which is known to have the form

$$w = f(\omega t - kx) + g(\omega t + kx). \tag{5.24}$$

By taking the second derivatives with respect to t and x respectively, we obtain

$$\frac{\partial^2 w}{\partial t^2} = \omega^2 \{ f''(\omega t - kx) + g''(\omega t + kx) \} \tag{5.25a}$$

$$\frac{\partial^2 w}{\partial x^2} = k^2 \{ f''(\omega t - kx) + g''(\omega t + kx) \}. \tag{5.25b}$$

Since the terms enclosed in { } are the same in both equations,

$$\frac{\partial^2 w}{\partial t^2} = \frac{\omega^2}{k^2} \frac{\partial^2 w}{\partial x^2}. \tag{5.26}$$

By comparing this with eqn (5.7), we see that

$$v = \frac{\omega}{k}. \tag{5.27}$$

In eqn (5.24), the first term represents a wave travelling to the right with a velocity of $v = \omega/k$, while the second term represents a wave travelling to the left. Note that these waves are not restricted to sine waves and can assume any arbitrary forms as long as f travels to the right and g to the left (see Fig. 5.4.)

Solutions for f and g are determined by the initial conditions (i.e. displacement w and $\partial w/\partial t$ at $t = 0$) and the boundary conditions. Two examples are shown in Fig. 5.5. In (a), the string is assumed to be infinitely long. When a section of the string is held in a triangular shape and then released, two triangular waves are created. Each wave has half the

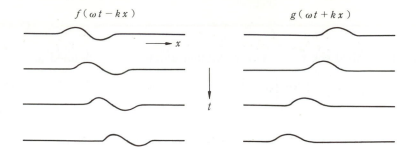

$f(\omega t - kx)$ $g(\omega t + kx)$

FIG. 5.4. Wave function f travels right and g travels left.

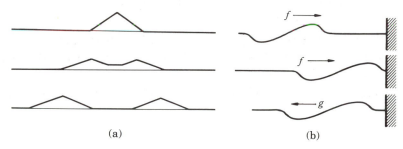

(a)   (b)

FIG. 5.5. Examples of initial and boundary conditions: (a) triangular wave released in an infinite string; (b) f, reflected at the fixed end, becomes g.

amplitude of the initial triangle, and the two waves travel in opposite directions.

In Fig. 5.5(b), the string is fixed at one end. When a wave f reaches the fixed end, it is reflected and becomes wave g. When both ends are fixed, one may mistakenly assume that the behaviour of the string will be complex, with waves f and g being reflected at both ends. Yet, as we have seen earlier, k assumes characteristic values only, and the solution is given by

$$w = \sum_{m=1}^{\infty} w_m = \sum_{m=1}^{\infty} (A_m \cos \omega_m t + B_m \sin \omega_m t) \sin\left(\frac{m\pi}{l}x\right). \quad (5.28)$$

If we take a term for a specific m value, we get

$$w_m = (A_m \cos \omega_m t + B_m \sin \omega_m t) \sin\left(\frac{m\pi}{l}x\right)$$

$$= \frac{1}{2} A_m \left\{ \sin\left[\omega_m t + \frac{m\pi}{l}x\right] - \sin\left(\omega_m t - \frac{m\pi}{l}x\right) \right\}$$

$$+ \frac{1}{2} B_m \left\{ \cos\left[\omega_m t - \frac{m\pi}{l}x\right] - \cos\left[\omega_m t + \frac{m\pi}{l}x\right] \right\}. \quad (5.29)$$

So

$$f_m = \frac{1}{2}\left\{B_m\cos\left(\omega_m t - \frac{m\pi}{l}x\right) - A_m\sin\left(\omega_m t - \frac{m\pi}{l}x\right)\right\} \qquad (5.30a)$$

$$g_m = \frac{1}{2}\left\{A_m\sin\left(\omega_m t + \frac{m\pi}{l}x\right) - B_m\cos\left(\omega_m t + \frac{m\pi}{l}x\right)\right\}. \qquad (5.30b)$$

5.4   Travelling-wave parameters and terminology

We have shown that the general solution for wave equation (5.7) is given by eqn (5.24), and that $f(x, t)$ and $g(x, t)$ are travelling waves. In this section, some of the parameters and terminology associated with travelling waves are briefly explained.

(1) *Phase*. The phase is the quantity expressed inside the parentheses for f and g in eqn (5.24). The meaning of phase will be clear when discussing sinusoidal waves in Section 5.12.

(2) *Phase velocity*. The relationship for t and x to make the phase zero is given by

$$\omega t \pm kx = 0. \qquad (5.31)$$

From this, taking account of eqns (5.13) and (5.8), we obtain

$$\frac{x}{t} = \pm\frac{\omega}{k} = \pm v = \pm\sqrt{\frac{T}{\rho}}. \qquad (5.32)$$

This quantity is called the phase velocity, which is denoted by v_{ph}. The phase velocity is the speed that is observed most easily by the travel of some point which has its phase remaining constant. The phase velocities for all other non-zero phases are the same, and are also given by eqn (5.32).

(3) *Wavelength λ and wavenumber k*. Figure 5.6 shows a simple wave travelling to the right. The length of one wave cycle is called the wavelength, expressed by λ. We can divide 2π by λ to obtain the wavenumber, expressed by k. The characteristic values of eqn (5.17) represent wavenumbers for standing waves.

(4) *Relation between phase velocity, wavelength and frequency*. The phase velocity of a travelling wave, given by eqn (5.32), is the product of wavelength λ and frequency ν, or

$$v_{\text{ph}} = \lambda\nu \qquad (5.33)$$

where $\nu = \omega/2\pi$.

(5) *Period*. The period is the time required for one wave cycle to be completed, and expressed by t_0 in this text. The period is the inverse of frequency:

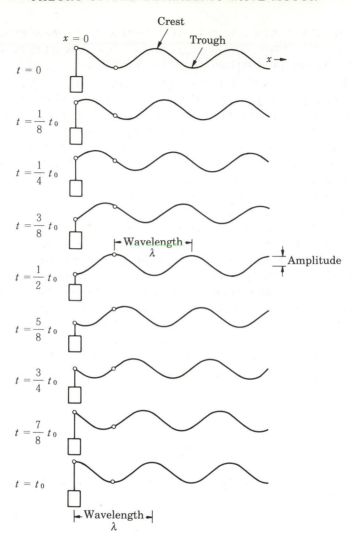

Fig. 5.6.   Time–space relation in a simple travelling wave.

$$t_0 = \frac{1}{\nu} = \frac{2\pi}{\omega}.$$ (5.34)

(6) *Dispersive and non-dispersive waves.* A wave for which the phase velocity and wavenumber (or wavelength) are independent of each other is called a non-dispersive wave. As in eqn (5.7), when the second derivatives with respect to time and position are proportional (and the proportionality constant is positive), the wave is non-dispersive. All other waves are disper-

sive. The velocity of a dispersive wave depends on its wavelength. In an ultrasonic-wave motor, which utilizes vibrations in a beam to be discussed later, the waves are dispersive.

(7) *Phase velocity and energy propagation velocity.* A travelling wave propagates energy as well as wave phases. The energy associated with a wave is the sum of its kinetic energy, which is due to the motion of points in the string or beam, and its potential energy (or strain energy), which is caused by the displacements of points against the 'pull' of tension (or elastic force). To understand the differences between the phase velocity and energy velocity, let us consider Fig. 5.7. This shows a local concentration of wave energy and is called a wave packet. When a wave packet travels, the energy associated with it travels at the same speed. However, as will be shown below in the examples of Fig. 5.8 or 5.10, the phase velocity of the wave is not always the same as the velocity at which the packet travels. When numerous wave packets are distributed evenly, they become superimposed to appear as a travelling wave, and individual wave packets disappear. As a result, only the phase velocity is visible. Yet the velocity of the energy transfer is regarded as the same as the velocity of a hidden wave packet. In this section we present the equations for the phase velocity and wave packet propagation velocity without detailed proofs.

Let us consider a wave equation of any form, which is a differential equation of time and displacement. We define $j\omega$ and jk as

$$\frac{\partial}{\partial t} = j\omega \qquad (5.35a)$$

$$\frac{\partial}{\partial x} = jk \qquad (5.35b)$$

and substitute these into the wave equation. From eqn (5.7), for example, we obtain

$$\omega^2 = v^2 k^2 \qquad (5.36)$$

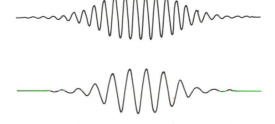

FIG. 5.7. Wave packets = concentrated wave energy.

or

$$\omega = \pm vk. \tag{5.37}$$

So the phase velocity v_{ph} is given by

$$v_{ph} = \frac{\omega}{k} = \pm v \tag{5.38}$$

where v is given by eqn (5.8).

On the other hand, the energy propagation velocity v_{eg} is given by

$$v_{eg} = \frac{\partial \omega}{\partial k} = \pm v. \tag{5.39}$$

As is obvious, in this case we have the non-dispersive relation

$$v_{ph} = v_{eg}. \tag{5.40}$$

Figure 5.8(a) shows a computer-derived solution of eqn (5.7). It shows how an initially placed wave packet splits into two smaller packets and then these travel in opposite directions. The phase velocity v_{ph} and wave packet velocity v_{eg} are equal in this case.

The reader should now refer to eqn (5.62) in Section 5.7 below, which is the wave equation for a beam. If we substitute the relationships given by eqns (5.35)(a) and (b) into this equation, we obtain

$$\omega^2 = \gamma k^4 \tag{5.41}$$

where

$$\gamma = EI/\rho A. \tag{5.42}$$

Therefore,

$$\omega = \pm \sqrt{\gamma} k^2 \tag{5.43}$$

and v_{ph} and v_{eg} are obtained as

$$v_{ph} = \frac{\omega}{k} = \pm \sqrt{\gamma} k \tag{5.44}$$

$$v_{eg} = \frac{\partial \omega}{\partial k} = \pm 2 \sqrt{\gamma} k. \tag{5.45}$$

From these we obtain

$$v_{eg} = 2v_{ph}. \tag{5.46}$$

That is, the energy propagation velocity is twice the phase velocity. This is illustrated in Fig. 5.8(b). Although we shall solve the beam's wave equation analytically in Section 5.9, the result obtained above was derived using an entirely different algorithm. BASIC programs are presented in Fig. 5.9.

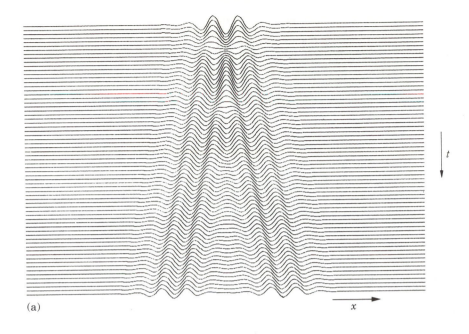

(a) x

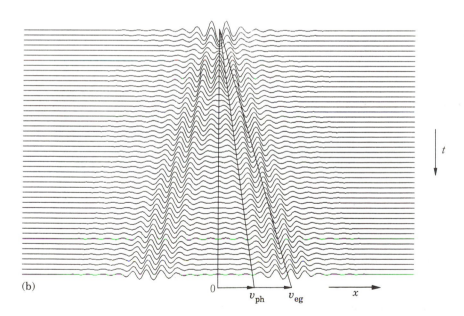

(b) 0

v_{ph} v_{eg} x

FIG. 5.8. (a) Propagation of a non-dispersive wave packet; (b) propagation of a dispersive wave packet governed by egn (5.62), with wave packet velocity twice the phase velocity.

```
100 REM 2nd order wave equation
110 CLS
120 DIM Y(301),V(301),D(301)
130 K=5 : DX=.007 : DT=.002 : PI=355/113
140 LINE(0,20)-(0,20)
150 FOR N=0 TO 300
160    Y(N)=COS(PI*N*.2)/(1+.00002*(N-150)^4)
170    X=2*N: Y=-Y(N)*7+20
180    LINE-(X,Y)
190 NEXT N
200 FOR M=0 TO 60
210    FOR N=1 TO 300
220       D(N)=Y(N+1)-2*Y(N)+Y(N-1)
230    NEXT N
240    FOR N=0 TO 300
250       V(N)=V(N)+K/DX^2*D(N)*DT
260       Y(N)=Y(N)+V(N)*DT
270    NEXT N
280    LINE(0,5*M+25)-(0,5*M+25)
290    FOR N=0 TO 300
300       X=2*N: Y=-Y(N)*7+25+5*M
310       LINE -(X,Y)
320    NEXT N
330 NEXT M
340 END
```

(a)

```
100 REM 4ht order wave equation
110 CLS: SCREEN 3
120 DIM Y(302),V(302),D(302)
130 K=.0005: DX=.01: DT=.002
140 LINE(0,20)-(0,20)
150 FOR N=0 TO 300
160    Y(N)=COS(3.1416*N*.2)/(1+.000005*(N-150)^4)
170    X=2*N: Y=-Y(N)*7+20
180    LINE-(X,Y)
190 NEXT N
200 FOR M=0 TO 60
210    FOR T=1 TO 2
220       FOR N=2 TO 300
230          D(N)=Y(N+2)-4*Y(N+1)+6*Y(N)-4*Y(N-1)+Y(N-2)
240       NEXT N
250       FOR N=2 TO 300
260          V(N)=V(N)-K/DX^4*D(N)*DT
270          Y(N)=Y(N)+V(N)*DT
280       NEXT N
290    NEXT T
300    LINE(0,5*M+25)-(0,5*M+25)
310    FOR N=0 TO 300
320       X=2*N: Y=-Y(N)*7+25+5*M
330       LINE -(X,Y)
340    NEXT N
350 NEXT M
360 END
```

(b)

Fɪɢ. 5.9. BASIC programs for solving wave equations: (a) for eqn (5.26); (b) for eqn (5.62).

A special wave equation is presented below, although unrelated to the ultrasonic motor. Suppose that a system exists in which the wave equation is given by

$$\frac{\partial^2 w}{\partial x \partial t} = Gw. \tag{5.47}$$

Then,

$$\omega k = -G \tag{5.48}$$

so the phase velocity and energy propagation velocity become

$$v_{\text{ph}} = \frac{\omega}{k} = -\frac{G}{k^2} \tag{5.49}$$

$$v_{\text{eg}} = \frac{\partial \omega}{\partial k} = +\frac{G}{k^2} \tag{5.50}$$

so

$$v_{\text{eg}} = -v_{\text{ph}}. \tag{5.51}$$

This means that the wave seems to move (i.e. movement of a phase) towards the vibrational source. This phenomenon is sometimes observed in sound waves travelling through a medium of weakly ionized plasma. Figure 5.10 shows a computer-derived illustration of such a wave. Although the wave packets travel to the right (making connections as they go along), the changing phase within the wave packet travels to the left. The behaviour of dispersive waves is fairly predictable as long as a sine waveform is maintained. If by some means the sine waveforms are disrupted, the wave will

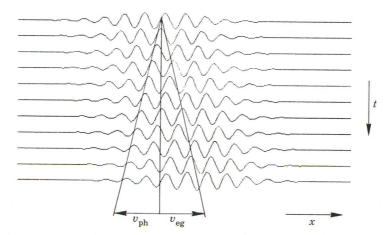

FIG. 5.10.   The phase and wave packet travel in opposite directions at the same velocity with a negative G.

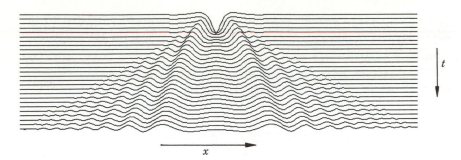

FIG. 5.11.   An impulse wave propagating outwards in a beam.

change its shape continually, making its behaviour complex.

Figure 5.11 shows the wave propagation in a beam, which satisfies eqn (5.64), after an initial impulse displacement is given. Wave components with shorter wavelengths travel faster, causing the wave to change its form as it progresses.

5.5   Characteristic impedance of waves

When a weight attached to a spring is pulled, then released, it will vibrate at a constant frequency (see Fig. 5.12). The vibration is caused by the mutual interaction of a 'restoring force' and an 'inertial force'. If the weight has mass M, and the spring constant is K, the angular frequency ω is given by

$$\omega = \sqrt{(K/M)}. \qquad (5.52)$$

In this case, K is the parameter for the restoring force, and M that of the inertial force. The vibration of a spring-mass system is locally contained

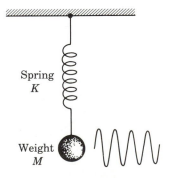

FIG. 5.12.   Weight–spring system.

and thus similar to the standing waves created in a stretched string. In the string, tension T represents the restoring force, while mass per unit length ρ represents the inertial force. Its frequency is given by eqn (5.20). The velocity of a wave travelling in a stretched string can also be expressed in terms of quantities T and ρ, as given in eqn (5.8), or

$$v = \sqrt{(T/\rho)}. \tag{5.53}$$

Thus v is an important parameter of a wave.

The characteristic impedance, also expressed in terms of T and ρ as

$$Z_0 = \sqrt{\rho T} \tag{5.54}$$

is another important parameter. We briefly discuss this subject below.

In Fig. 5.13, a stretched string is attached at its left-hand end ($x = 0$) to a vibration source which vibrates vertically in simple harmonic oscillation. In the equilibrium state of (a), tension T is applied in the x-direction. When the string is vibrating, as in (b), the force exerted on the drive source by tension T has a y-component F_y given by

$$F_y = T \sin \theta. \tag{5.55}$$

Since θ is assumed to be small,

$$F_y = T \sin \theta \simeq T \tan \theta \tag{5.56}$$

$$= T \frac{\partial w}{\partial x}. \tag{5.57}$$

If a wave $f(\omega t - kx)$, travelling in the positive x-direction, is given by either

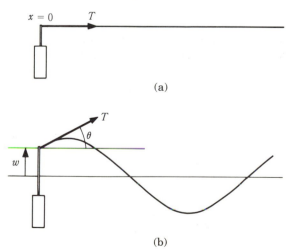

(a)

(b)

Fɪɢ. 5.13. Transverse wave created in a stretched string by the vibrator: (a) equilibrium state; (b) vibrating state.

$$\sin(\omega t - kx) \quad \text{or} \quad \cos(\omega t - kx) \tag{5.58}$$

the following relation can be established:

$$F_y = -\frac{T}{v}\frac{\partial w}{\partial t} \tag{5.59}$$

where F_y is the wave-generating force, T/v the impedance ($-Z_0$), and $\partial w/\partial t$ the velocity of the string in the transverse direction. The impedance in this case is a mechanical parameter and is analogous to the impedance in an electrical circuit, where

$$V = Z \cdot I \tag{5.60}$$

V being the voltage, Z the impedance and I the current.

Equation (5.54) can be derived from eqn (5.53) as follows:

$$Z_0 = \frac{T}{v} = \frac{T}{\sqrt{(T/\rho)}} = \sqrt{\rho T}. \tag{5.61}$$

The meaning of the characteristic impedance is illustrated in Fig. 5.14. The mechanical impedance Z_0 of the string, viewed from the vibrator, is given as above by eqn (5.61). However, the impedance seen towards the right at any point must be the same, as the string is assumed to stretch out to infinity and no reflected waves travelling in the opposite direction exist. Simple logic demonstrates that such impedance is the same as Z_0, the characteristic impedance.

5.6 Creating one-way travelling waves

A wave travelling in one direction on an infinite string was described earlier in this chapter. Since infinite strings do not actually exist, certain arrangements are necessary to create such waves on a string of finite length. Figure 5.15 shows two methods. In either method, two vibrators are used: one generates the vibrations, while the other absorbs those vibrations generated. Langevin vibrators are used for both vibrators: the vibration generator acts

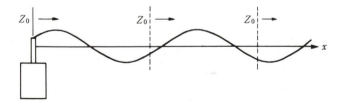

FIG. 5.14. A string infinitely stretched to the right has characteristic impedance Z_0 at any point when viewed to the right.

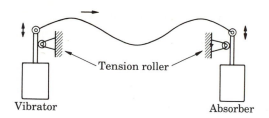

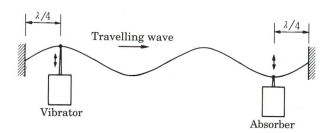

FIG. 5.15. Creating travelling waves with a vibrator and absorber.

as the drive source, referred to in the previous section, while the absorber acts as an electric generator, which is used to match the mechanical impedance to the characteristic impedance (by connecting a load with suitable impedance). It is possible to create an 'infinite' string, in effect, by using such arrangements.

5.7 Wave equation for beams

In the ultrasonic-wave motor, vibrations in an elastic body are used. The wave equations for elastic bodies are more complex than those for stretched

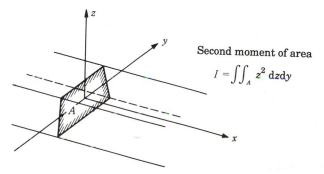

FIG. 5.16. Second moment of area about the *y*-axis (not limited to rectangular cross-sections).

strings. Even with simplifications and approximations, flexural vibration is expressed by the fourth-order partial differential equation

$$\frac{\partial^2 w}{\partial t^2} = -\frac{EI}{\rho A}\frac{\partial^4 w}{\partial x^4} \tag{5.62}$$

where E is the Young's modulus, ρ the mass per unit volume, A the cross-sectional area, and I the second moment of area about the y-axis (passing through the neutral plane), defined, as illustrated in Fig. 5.16, by

$$I = \iint_A z^2\,dz\,dy. \tag{5.63}$$

We shall first derive eqn (5.62). As shown in Fig. 5.17, a plane exists in the beam in which only bending (but no elongation) occurs. This is called the neutral plane. We shall let the neutral plane's initial (undeformed) position correspond to $z = 0$. The bending is assumed to take place parallel to the zx-plane, and the displacement in the z-direction is given by $w(x, t)$. Letting $S(x, t)$ be the shear force acting along a beam cross-section, and $M(x, t)$ be the bending moment, the equation of motion of a segment with length Δx is given by

$$\rho A \Delta x \frac{\partial^2 w}{\partial t^2} = \frac{\partial S}{\partial x}\Delta x \tag{5.64}$$

$$\rho A \frac{\partial^2 w}{\partial t^2} = \frac{\partial S}{\partial x}. \tag{5.65}$$

Shear and moment have the following relation:

$$S = \frac{\partial M}{\partial x}. \tag{5.66}$$

This can be derived as follows. The equation of rotational motion about the centre of gravity G (Fig. 5.18) is given by

$$I_0\frac{d^2\theta}{dt^2} = -\left(S + \frac{\partial S}{\partial x}\Delta x\right)\frac{\Delta x}{2} - S\frac{\Delta x}{2} + \left(M + \frac{\partial M}{\partial x}\Delta x\right) - M. \tag{5.67}$$

The second moment of area I_0 of segment Δx is proportional to $(\Delta x)^2$. If

Fig. 5.17.   Flexural wave in a beam and the neutral plane.

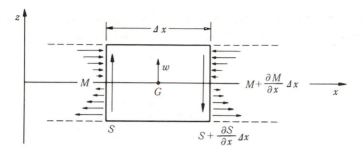

Fig. 5.18. Vertical motion of a beam segment's centre of gravity G.

Δx is chosen sufficiently small, the terms including $(\Delta x)^2$ can be neglected. Thus

$$0 = -S\Delta x + \frac{\partial M}{\partial x}\Delta x \qquad (5.68)$$

from which eqn (5.66) follows.

Substituting eqn (5.66) into eqn (5.65), we obtain

$$\rho A \frac{\partial^2 w}{\partial t^2} = \frac{\partial^2 M}{\partial x^2}. \qquad (5.69)$$

Next, we examine the radius of curvature of the neutral plane. We shall assume that a small segment PQ along the neutral plane bends in a circular arc, with O_0 as the circle's centre (see Fig. 5.19). The tangent lines, PK and QL form angles θ and $\theta + d\theta$, respectively, with the x-axis. It can be seen that $d\theta$ is also the angle PO_0Q. Then the following relations are obtained from the diagram:

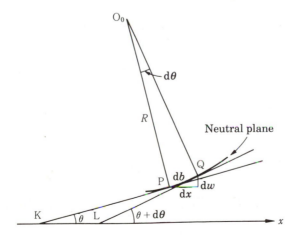

Fig. 5.19. Radius of curvature of the neutral plane.

$$db = \pm R \, d\theta \qquad (5.70)$$

$$\frac{dw}{dx} = \tan \theta. \qquad (5.71)$$

Differentiating eqn (5.71) with respect to x, we obtain

$$\frac{d^2 w}{dx^2} = \frac{1}{\cos \theta^2} \frac{d\theta}{dx}. \qquad (5.72)$$

Since $\theta \simeq 0$, we can assume that $\cos \theta \simeq 1$. Furthermore, substituting eqn (5.70) into (5.72), we obtain

$$\frac{d^2 w}{dx^2} = \pm \frac{1}{R} \frac{db}{dx}. \qquad (5.73)$$

Since $db/dx \simeq 1$ because $\theta \simeq 0$, we obtain

$$\frac{d^2 w}{dx^2} = \pm \frac{1}{R}. \qquad (5.74)$$

Next we consider the bending strain. In Fig. 5.20, the strain above or below the neutral plane is given by

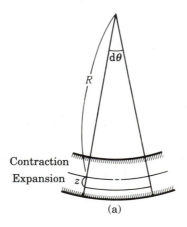

(a)

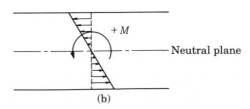

(b)

Fɪɢ. 5.20. Deformation and stress distribution in a bent beam: (a) longitudinal deformation; (b) stress distribution and resultant bending moment.

$$h = \frac{z}{R} \approx \pm z \frac{d^2w}{dx^2}. \tag{5.75}$$

This strain results in a moment M (see Fig. 5.20b). The stress T at a point caused by deformation in the axial direction is given, using Young's modulus E, as

$$T = Eh = \pm Ez \frac{d^2w}{dx^2}. \tag{5.76}$$

The moment, when the sign is defined as in Fig. 5.20b (counterclockwise is regarded as positive), is given by

$$M = \iint_A Tz \, dx \, dy = +E \frac{d^2w}{dx^2} \iint_A z^2 \, dx \, dy. \tag{5.77}$$

Substituting I, defined in eqn (5.63), we obtain

$$M = EI \frac{d^2w}{dx^2}. \tag{5.78}$$

If we substitute this into eqn (5.69), we obtain

$$\rho A \frac{\partial^2 w}{\partial t^2} = -EI \frac{\partial^4 w}{\partial x^4} \tag{5.79}$$

and eqn (5.62) immediately follows.

5.8 Precise wave equation

Equation (5.62), which governs the flexural vibrations of a beam, was derived by making several assumptions and approximations. In this section, equations that are more accurate are presented for interested readers, before we investigate the beam's vibrational behaviour in the following section.

(1) If we take into account the rotational energy of each point in the beam, the wave equation is

$$\frac{EI}{\rho A} \frac{\partial^4 w}{\partial x^4} + \frac{\partial^2 w}{\partial t^2} - \frac{I}{A} \frac{\partial^4 w}{\partial x^2 \partial t^2} = 0. \tag{5.80}$$

(2) Further, taking into account the shear strain results in

$$\frac{EI}{\rho A} \frac{\partial^4 w}{\partial x^4} + \frac{\partial^2 w}{\partial t^2} - \frac{I}{A}\left(1 + \frac{E}{\mu}\varsigma\right)\frac{\partial^4 w}{\partial x^2 \partial t^2} + \frac{\rho \varsigma I}{\mu A} \frac{\partial^4 w}{\partial t^4} = 0 \tag{5.81}$$

where μ is the shear modulus (modulus of rigidity), and ς a correction factor, equal to 9/8 for rectangular cross-sections.

Flexural waves are the superimposed results of both transverse and longitudinal waves, and their behaviour can be completely described by eqn (5.81). Actually, this equation still does not suffice for accurate analysis of the ultrasonic motor, since part of the vibration energy is transmitted to the rotor to be converted to mechanical output. Moreover, considerable energy is dissipated as heat loss through friction. These effects are not taken into consideration in the equations above.

The comb-teeth section poses another problem. The precise relation of eqn (5.81) is based upon the assumption that the beam is uniform and therefore that the Young's modulus is constant throughout the beam. A thorough theoretical analysis of the actual motor with comb-teeth on the stator is therefore extremely difficult.

Instead of the complex equations presented here, we shall use the simpler equation (5.62) for our later investigations. Since this equation describes only transverse motions of a point on a beam, our approach is to study the basic vibrational behaviour of the beam using this equation, then consider longitudinal effects using approximate methods. The effect of the contact with a rotor will be treated by a different approach in Chapter 6.

5.9   Vibrations in a beam

Unlike stretched strings, the analysis of beam vibration is complex because its equation includes a fourth-order derivative with respect to x. If we assume as before that the beam vibrates in simple harmonic motion with a frequency of $\omega/2\pi$, the displacement w can be expressed, using undetermined constants P and Q, in the form

$$w = (P\cos \omega t + Q\sin \omega t)q(x).$$
(5.82)

Then

$$\frac{d^4 q}{dx^4} = \beta^4 q$$
(5.83)

where

$$\beta^4 = \frac{\omega^2 \rho A}{EI}.$$
(5.84)

The general solution for this is given by

$$q = C_1 \cosh \beta x + C_2 \sinh \beta x + C_3 \cos \beta x + C_4 \sin \beta x.$$
(5.85)

We shall examine two specific cases.

5.9.1 Both beam ends supported

When the beam is simply supported at $x = 0$ and $x = 1$, the boundary conditions are given as

$$q = 0, \quad \frac{d^2q}{dx^2} = 0 \tag{5.86}$$

From the condition at $x = 0$, we see that $C_1 = C_3 = 0$, and from the condition at $x = l$ we obtain

$$C_2 \sinh \beta l + C_4 \sin \beta l = 0 \tag{5.87a}$$

$$C_2 \sinh \beta l - C_4 \sin \beta l = 0. \tag{5.87b}$$

A non-zero solution for q is obtained when

$$\sinh \beta l \sin \beta l = 0. \tag{5.88}$$

We see that a solution exists when $C_2 = 0$, so

$$\beta l = n\pi \quad (n = 1, 2, \ldots) \tag{5.89}$$

The solution for q is thus given by

$$q = C_{4,n} \sin \frac{n\pi}{l} x. \tag{5.90}$$

In eqn (5.89), β is the eigenvalue of eqn (5.82), and eqn (5.90) is its characteristic equation. The characteristic frequency can be derived from eqns (5.84) and (5.89), as

$$v_n = \frac{\omega_n}{2\pi} = \frac{1}{2\pi}\sqrt{\frac{EI}{\rho A}} \beta^2 = \frac{1}{2\pi}\sqrt{\frac{EI}{A\rho}} \frac{n^2\pi^2}{l^2}. \tag{5.91}$$

Thus the vibration of a beam simply supported at both ends can be expressed by the following equation:

$$w_n = (P_n \cos \omega_n t + Q_n \sin \omega_n t) \sin \frac{n\pi}{l} x. \tag{5.92}$$

Although this equation is similar to the one for vibrations on a stretched string, there are some differences, which are listed below.

(1) In the case of string vibrations, the characteristic frequencies are integer (n) multiples of the fundamental frequency, whereas in beam vibrations, they are integer-squared (n^2) multiples of the fundamental frequency.

(2) In string vibrations, the velocity for travelling waves was unrelated to the frequency, whereas in beams it is proportional to the mode n as given by

$$\text{velocity of travelling wave} = \frac{\omega_n}{k_n} = \sqrt{\frac{EI}{\rho A}} \; \frac{n\pi}{l} \qquad (5.93)$$

because we have the following relation:

$$\frac{1}{k_n} = \frac{\lambda_n}{2\pi} = \frac{2l/n}{2\pi} = \frac{l}{n\pi}. \qquad (5.94)$$

(3)   A general solution, such as eqn (5.24), does not exist for flexural waves in beams.

Radio waves and sound waves propagating through air have similar characteristics to those of string vibrations, since the equations can be expressed in terms of the second derivative with respect to displacement. Not only are beam vibrations more complex, but in ultrasonic motors the problem of friction exists. Thus, to appreciate the problems involved in ultrasonic motor design, a careful study of beam vibration phenomena is required.

5.9.2   Vibrations of a cantilevered beam

In Chapter 4, where we discussed the wedge-type motor, the characteristic frequency of the vibrator piece was given by eqn (4.11). In this section, we shall derive this equation.

The boundary conditions for a beam of length l, which is fixed at one end $x = 0$, are given by

$$q = 0, \quad \mathrm{d}q/\mathrm{d}x = 0 \qquad\qquad \text{at fixed end } x = 0 \qquad (5.95\text{a})$$

$$\mathrm{d}^2q/\mathrm{d}x^2 = 0, \quad \mathrm{d}^3q/\mathrm{d}x^3 = 0 \quad \text{at free end } x = l. \qquad (5.95\text{b})$$

Applying the conditions at $x = 0$ to eqn (5.85), we obtain

$$C_1 + C_3 = 0, \quad C_2 + C_4 = 0. \qquad (5.96)$$

So eqn (5.85) becomes

$$q = C_1(\cosh \beta x - \cos \beta x) + C_2(\sinh \beta x - \sin \beta x). \qquad (5.97)$$

Applying the conditions at $x = l$, we obtain the two equations

$$\left. \begin{array}{l} C_1(\cosh \beta l + \cos \beta l) + C_2(\sinh \beta l + \sin \beta l) = 0 \\ C_1(\sinh \beta l - \sin \beta l) + C_2(\cosh \beta l + \cos \beta l) = 0 \end{array} \right\} \qquad (5.98)$$

The first equation can be written as

$$C_1(\cosh \beta l + \cos \beta l) = - C_2(\sinh \beta l + \sin \beta l). \qquad (5.99)$$

By equating both sides of this equation with G, we get

$$C_1 = \frac{G}{\cosh \beta l + \cos \beta l} \qquad (5.100\text{a})$$

$$C_2 = -\frac{G}{\sinh \beta l + \sin \beta l}.$$ (5.100b)

Then eqn (5.97) can be expressed as follows:

$$q = G\left(\frac{\cosh \beta x - \cos \beta x}{\cosh \beta l + \cos \beta l} - \frac{\sinh \beta x - \sin \beta x}{\sinh \beta l + \sin \beta l}\right).$$ (5.101)

By eliminating C_1 and C_2 from eqns (5.98), we obtain the following equation for βl:

$$1 + \cosh \beta l \cos \beta l = 0.$$ (5.102)

There are an infinite number of solutions to this equation, of which the smallest three are

$$\beta_1 l = 0.60\pi, \quad \beta_2 l = 1.49\pi, \quad \beta_3 l = 2.50\pi.$$

From the characteristic value $\beta_n = a_n \pi/l$, we obtain the characteristic frequency as

$$\nu_n = \frac{\omega_n}{2\pi} = \frac{a_n^2 \pi^2}{2\pi l^2}\sqrt{\frac{EI}{A\rho}}.$$ (5.103)

If we substitute $a_1 = 0.6$ in this equation, we obtain eqn (4.11).

5.10   Controlling travelling waves in a linear motor

In order to create a linear ultrasonic-wave motor, it is necessary to generate a travelling wave which moves in only one direction. A common method is to use two Langevin vibrators, as shown in Fig. 5.21: one for creating

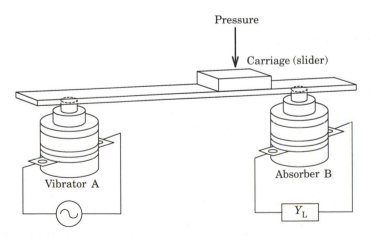

FIG. 5.21.   Vibrator and absorber in a linear ultrasonic wave motor.

oscillations and the other for absorbing them. We shall let A be the vibrator
and B be the absorber. B absorbs the oscillations to generate electricity,
which is consumed in the connected load circuit. The impedance of this circuit must be adjusted (or matched) properly so as to equate the mechanical
impedance of the absorber B to the characteristic impedance Z_0 of the
beam. Otherwise the oscillations will not be totally absorbed, some will be
reflected back and a standing wave may form. We have already discussed
the concept of the characteristic impedance associated with a travelling
wave in Section 5.5. The expression for the approximate value of Z_0 for
a beam with a quadrilateral cross-section is known to be given by

$$Z_0 = \frac{Edh^3k^3}{24\pi\nu} \text{ Nms}^{-1} \tag{5.104}$$

where E is the Young's modulus (Nm^{-2}), d and h are the width and height
respectively of the cross-section (m), k is the wavenumber ($2\pi/\lambda$) (m^{-1}),
and ν is the vibration frequency (s^{-1}).

Note however that the effects of longitudinal vibrations taking place outside the neutral plane are not considered in this equation. The actual
characteristic impedance will thus be slightly larger than the calculated
value obtained from this equation. The impedance expressed by eqn (5.104)
is that of a mechanical system, and its dimension is N s m^{-1}. The basic
idea for effecting the impedance matching of the absorber load circuit is
schematically shown in Fig. 5.22. If the ratio of the larger section of the
horn to the smaller section is n, the force applied to the absorber B is
amplified n times by the expansions and contractions in the horn before
being transmitted to the piezoelectric ceramic, while the transverse velocity
at the connecting part of the beam and the horn is reduced by $1/n$ at the
ceramic. The piezoelectric ceramic produces a voltage which is equal to the
applied force multiplied by A^{-1} (A is the force factor here). In other
words, a voltage which is equal to n/A times the force applied to the end
of the absorber is applied to the load circuit. This process is analogous to
the principle of impedance transformation by a transformer, and the mechanical impedance Z, viewed from the terminal side, is given by

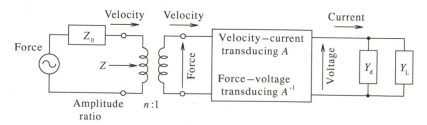

FIG. 5.22. Impedance matching $Z = Z_0$: Z_0, characteristic impedance; Y_d,
vibrator's blocked admittance $= j\omega C_d$; Y_L, load admittance; A, force factor;
n, horn's vibration amplification factor.

$$Z = \left(\frac{A}{n}\right)^2 \frac{1}{Y_d + Y_L}. \tag{5.105}$$

For the load Y_L, a parallel circuit consisting of an inductor and a resistor is commonly used. The impedance matching is achieved by adjusting Y_L so that $Z = Z_0$ is obtained.

In this model we assume that part of the energy supplied by the oscillator to the beam is spent in driving the slider (or carriage) pressed on the beam without essentially influencing the oscillation behaviour. Therefore the absorber receives the supplied energy minus the mechanical output and friction loss at the contact surface.

5.11   Treatment of travelling waves for rotary motors

Instead of a beam, we now consider an elastic-body ring. Figure 5.23 shows a simple model for treating flexural waves that propagate on a ring. Consider first an infinite beam which carries flexural waves. A ring is formed by cutting a length equal to the wavelength times m, then bending the section and joining the two ends together. In Section 5.9.1 we stated that, for a beam, there are no general solutions which can be expressed in the form of eqn (5.24). There is however a sinusoidal vibration which can be expressed as

$$w_m = C_m \sin\left[\omega_m t - \frac{2m\pi}{l}x + \phi_m\right]$$

$$+ D_m \sin\left[\omega_m t + \frac{2m\pi}{l}x + \varphi_m\right] \tag{5.106}$$

$W_n = (P_n \cos\omega_n t +$

$\quad Q_n \sin\omega_n t) \sin\frac{n\pi}{\ell} x$

where ϕ_m and φ_m are appropriate phase angles, and l is equal to the wavelength λ times m. By comparing this equation with eqn (5.92), we can $n = 2m$

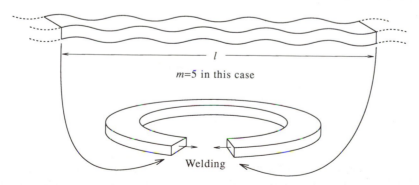

l

$m=5$ in this case

Welding

FIG. 5.23.   Stator model for a rotary motor (ends are welded together).

infer the following relation between n, the vibrational mode number, and m, the number of wave cycles present on the ring (which also corresponds to the number of pole pairs and should not be confused with m in Fig. 5.3 for string vibrations):

$$n = 2m. \tag{5.107}$$

From this equation we see that $n = 1$ corresponds to one-half of a wave cycle on the ring (as we saw in Section 5.9.2) and so is not possible. This argument applies for $n = 3, 5, 7, \ldots$ as well; in general, n cannot be an odd integer. The characteristic frequency in this case is given by substituting the value for n obtained above into eqn (5.91).

If, instead of eqn (5.106), we use polar coordinates, the neutral plane's displacement w_m in the z-direction is given by

$$w_m = C(r) \sin(\omega_m t - m\theta + \phi_m) + D(r) \sin(\omega_m t + m\theta + \psi_m). \tag{5.108}$$

Figure 5.24 shows a ring carrying a ninth-mode wave (i.e. nine wave cycles along the ring's perimeter). As this shows, in an actual motor, the amplitude increases towards the outer perimeter. In this text, however, the flexural wave on an elastic-body ring is given a semiquantitative, semiqualitative treatment using the approximate solution derived with orthogonal coordinates. Note that eqn (5.108) contains two wave components: one propagating clockwise, the other counterclockwise. As we shall see in the following chapters, only one of these waves is excited to drive a motor.

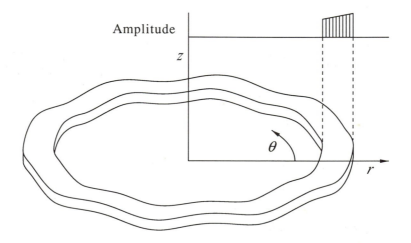

FIG. 5.24.   Flexural wave on an elastic ring.

5.12 Elliptical motion at the beam's surface

Figure 5.25 shows how points on the beam's top surface rotate counter-clockwise in elliptical motion as the flexural wave travels in the → direction. We shall closely examine this elliptical motion, since this is what drives the motor. Assume that the vertical displacement of the neutral plane is given by

$$w = \xi_0 \sin(\omega t - kx). \qquad (5.109)$$

As explained in Fig. 5.25, the displacement u in the x direction equals the product of the slope of the neutral plane and half the beam height. That is,

$$u = (k\xi_0 h/2) \cos(\omega t - kx). \qquad (5.110)$$

Since $k = 2\pi/\lambda$, we obtain

$$u = (\pi\xi_0 h/\lambda) \cos(\omega t - kx). \qquad (5.111)$$

We can see from this that the ratio of the minor to the major axis of an ellipse is given by $\pi h/\lambda$. For $\lambda = 40\,\text{mm}$ and $h = 2\,\text{mm}$, this value is equal to 0.157. For an amplitude of $1\,\mu\text{m}$, this results in a horizontal

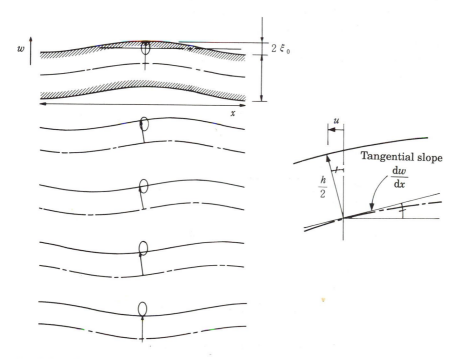

FIG. 5.25. Points at the beam's surface rotate counterclockwise in an ellipse as travelling waves move right. The ellipse's minor-axis/major-axis ratio is $\pi h/\lambda$.

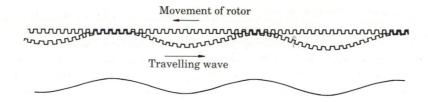

FIG. 5.26. Rotor movement created by travelling wave when rotor and stator are engaged with gear teeth.

movement of $0.157\,\mu$m. If we assume that the stator and rotor rings have gear teeth so that they can engage one another, as shown in Fig. 5.26, the rotor will rotate as travelling waves are generated. The direction of the rotation will be opposite to that of the wave's movement.

Currently developed ultrasonic-wave motors use friction instead of meshed gear teeth to transmit forces and drive the ring rotor. As was hinted in Chapter 2, the frictional behaviour in an ultrasonic-wave motor is complex, requiring detailed theoretical and experimental investigations.

Now let us derive the tangential velocity of each point at the beam surface, by differentiating eqn (5.111) with respect to t:

$$\frac{\mathrm{d}u}{\mathrm{d}t} = \frac{-\omega\pi\xi_0 h}{\lambda} \sin(\omega t - kx). \tag{5.112}$$

By comparing with eqn (5.109), it is seen that the speed is zero at the node of the wave, and maximum at the crest or trough. Figure 5.27 shows how the beam surface expands and contracts as the wave proceeds. This

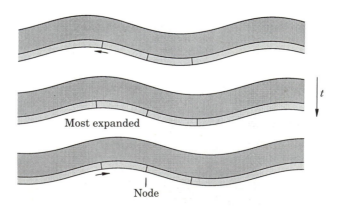

FIG. 5.27. Illustration of the phase relation between the piezoeletric ceramic's expansion/contraction and the beam's flexural motion.

gives an important clue for determining the force factor of the piezoelectric element when used in an ultrasonic-wave motor in Chapter 7.

Up to this point, we have assumed that the A and B phases were identical, except for the $\pi/2$ phase difference in position and time. However, this may not necessarily be the case. Thus, let ξ_A and ξ_B be the amplitudes (at the neutral plane) of the A and B phases, respectively. Then we have

$$w = \xi_A \sin \omega t \sin kx + \xi_B \cos \omega t \cos kx \qquad (5.113)$$

$$u = -\frac{h}{2} \frac{\partial w}{\partial x}. \qquad (5.114)$$

The tangential velocity v_s at the upper surface is given by

$$v_s = \frac{\partial u}{\partial t} = -\frac{h}{2} \frac{\partial^2 w}{\partial x \partial t} \qquad (5.115)$$

$$= \frac{-hk\omega}{2}(\xi_A \cos \omega t \cos kx + \xi_B \sin \omega t \sin kx). \qquad (5.116)$$

To find the position of the crest, we let $\partial w/\partial x = 0$. Thus

$$\xi_A \sin \omega t \cos kx = \xi_B \cos \omega t \sin kx. \qquad (5.117)$$

From this equation, we obtain the following:

$$\cos kx = \frac{(\xi_B/\xi_A) \cot \omega t}{\sqrt{[1 + (\xi_B/\xi_A)^2 \cot^2 \omega t]}} \qquad (5.118)$$

$$\sin kx = \frac{1}{\sqrt{[1 + (\xi_B/\xi_A)^2 \cot^2 \omega t]}}. \qquad (5.119)$$

Substituting these terms into eqn (5.116), we obtain the tangential velocity at the crest:

$$[v_s]_{top} = \frac{-(hk\omega/2)\xi_A \xi_B}{\sqrt{(\xi_A^2 \sin^2 \omega t + \xi_B^2 \cos^2 \omega t)}} \qquad (5.120)$$

$$[v_s]_{top} = \frac{-(h\pi\omega)\xi_A \xi_B}{\lambda \sqrt{(\xi_A^2 \sin^2 \omega t + \xi_B^2 \cos^2 \omega t)}}. \qquad (5.121)$$

From this equation, the following observations can be made.

(1) *When $\xi_A = \xi_B = \xi_0$,*

$$[v_s]_{top} = \frac{-h\pi\omega\xi_0}{\lambda} = \frac{-\left(\dfrac{h}{2}\right)\xi_0\omega}{(\lambda/2\pi)}. \qquad (5.122)$$

This equation can be intrepreted as follows. A diagram of an L-shaped lever is shown in Fig. 5.28. The length of the long arm is equal to the wavelength

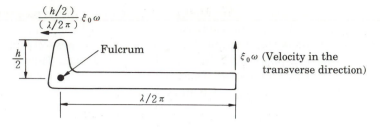

FIG. 5.28.   Each comb-tooth as an L-shaped lever.

divided by 2π, while the short arm has length $h/2$. The velocity at the crest is that obtained at the end of the short arm, when a velocity of $\xi_0\omega$ is given to the tip of the long arm. The quantity $\xi_0\omega$ can be considered as the equivalent velocity of the beam's longitudinal vibration.

This analogy applies directly to the actual motor, where the comb-teeth are considered to be the short arms of the lever. The comb-teeth are thus regarded as a row of levers, as shown in Fig. 5.29.

(2)   *When $\xi_A \neq \xi_B$.* In an actual motor, it is unlikely for the two phases to have identical amplitudes. How then should the model given above to explain the force transmission mechanism be modified to satisfy actual conditions? By comparing eqns (5.121) and (5.122), we can see that the solution is to substitute the term $\xi_A\xi_B/\sqrt(\xi_A^2 \sin^2\omega t + \xi_B^2 \cos^2\omega t)$, which includes the vibrator's components, for ξ_0. It can be shown that both the wave height and the tangential velocity at the crest can be expressed in terms of the excitation frequency multiplied by 2. By substituting eqns (5.118) and (5.119) into eqn (5.113), we obtain the change in the wave height as

$$w_{\text{top}} = \sqrt(\xi_A^2 \sin^2\omega t + \xi_B^2 \cos^2\omega t) \tag{5.123}$$

which can be rewritten as

$$w_{\text{top}} = \sqrt[\xi_A^2 \sin^2\omega t + \xi_A^2 \cos^2\omega t - (\xi_A^2 - \xi_B^2)\cos^2\omega t]$$

$$= \xi_A\sqrt{\left(1 - \frac{\xi_A^2 - \xi_B^2}{\xi_A^2}\cos^2\omega t\right)}.$$

Assuming that

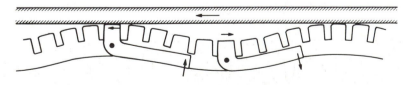

FIG. 5.29.   The 'lever' principle in the ultrasonic wave motor: the beam's transverse waves are converted to motion in the tangential direction.

$$\xi_A^2 \gg |\xi_A^2 - \xi_B^2| \qquad (5.124)$$

we obtain

$$w_{\text{top}} \simeq \xi_A \left\{ 1 - \frac{1}{2} \left(\frac{\xi_A^2 - \xi_B^2}{\xi_A^2} \right) \cos^2 \omega t \right\}$$

$$= \xi_A \left\{ 1 - \frac{1}{4} \left(\frac{\xi_A^2 - \xi_B^2}{\xi_A^2} \right) (1 + \cos 2\omega t) \right\}. \qquad (5.125)$$

In a similar manner, eqn (5.120) becomes

$$[v_s]_{\text{top}} = \frac{-(hk\omega/2)\xi_B}{\sqrt{\left\{ \sin^2 \omega t + \cos^2 \omega t - \left(\frac{\xi_A^2 - \xi_B^2}{\xi_A^2} \right) \cos^2 \omega t \right\}}} \qquad (5.126)$$

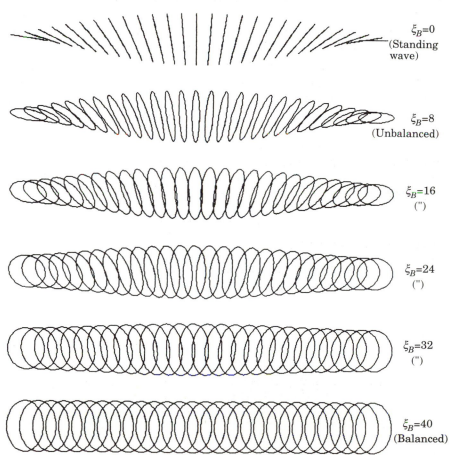

$\xi_B = 0$
(Standing wave)

$\xi_B = 8$
(Unbalanced)

$\xi_B = 16$
(")

$\xi_B = 24$
(")

$\xi_B = 32$
(")

$\xi_B = 40$
(Balanced)

FIG. 5.30. Motion of points under unbalanced and balanced conditions.

$$\simeq \frac{(hk\omega/2)\xi_B}{1 - \frac{1}{2}\{(\xi_A^2 - \xi_B^2)/\xi_A^2)\cos^2\omega t\}}$$

$$\simeq \frac{hk\omega\xi_B}{2}\left[1 + \frac{\xi_A^2 - \xi_B^2}{2\xi_A^2}\cos^2\omega t\right]$$

$$= \frac{hk\omega\xi_B}{2}\left\{1 + \frac{\xi_A^2 - \xi_B^2}{4\xi_A^2}(1 + \cos 2\omega t)\right\}. \tag{5.127}$$

It is interesting to note the following relation

$$w_{\text{top}}[v_s]_{\text{top}} = -\xi_A\xi_B(hk\omega/2). \tag{5.128}$$

In other words, the product of the wave height and tangential velocity is a constant. What are the motions taken by the points at the beam's upper surface when such imbalances exist? Figure 5.30 shows the results obtained with a BASIC program. In this example, the amplitude ξ_A is fixed at 20,

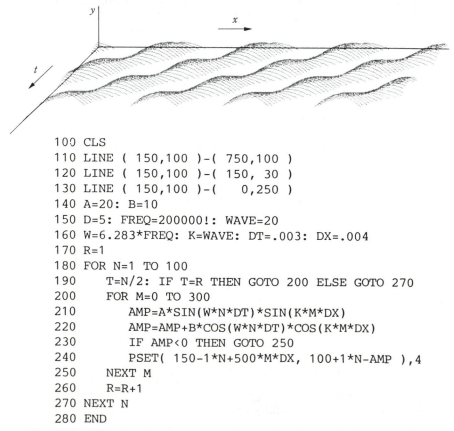

```
100 CLS
110 LINE ( 150,100 )-( 750,100 )
120 LINE ( 150,100 )-( 150, 30 )
130 LINE ( 150,100 )-(   0,250 )
140 A=20: B=10
150 D=5: FREQ=200000!: WAVE=20
160 W=6.283*FREQ: K=WAVE: DT=.003: DX=.004
170 R=1
180 FOR N=1 TO 100
190     T=N/2: IF T=R THEN GOTO 200 ELSE GOTO 270
200     FOR M=0 TO 300
210         AMP=A*SIN(W*N*DT)*SIN(K*M*DX)
220         AMP=AMP+B*COS(W*N*DT)*COS(K*M*DX)
230         IF AMP<0 THEN GOTO 250
240         PSET( 150-1*N+500*M*DX, 100+1*N-AMP ),4
250     NEXT M
260     R=R+1
270 NEXT N
280 END
```

FIG. 5.31.   Wave under an unbalanced condition, and a BASIC program.

while ξ_B is varied between 0 and 20. When $\xi_B = 0$ (i.e. standing wave), the points move in straight lines. When $\xi_B \neq 0$ and $\xi_B \neq \xi_A$, elliptical motion is observed with the amplitudes varying according to the point's location, and the major axes are tilted.

The illustration in Fig. 5.31 shows the theoretically derived wave configuration under an unbalanced condition. Two wave ripples can be observed within one cycle. The BASIC program that was written to derive the wave configuration is also given.

5.13 Losses at the contact surface

The most critical part in the ultrasonic-wave motor is where the rotor and stator come into contact to transmit torque and thrust. So far we have examined some basic aspects of beam vibrations. In this section we investigate the problem of losses in the linear motor. To start with, we assume that forces are being transmitted from stator to rotor extremely efficiently, with no losses occurring due to friction. Let $f(x, t)$ be the tangential force acting on the contact surface at each point. Then the work P_{IN} done on the rotor per unit time is given by the following surface integral:

$$P_{IN} = \int_s \frac{1}{t_0} \int_t^{t+t_0} f \frac{du}{dt} \, dt \, dS \quad (dS = dx \, dy) \tag{5.129}$$

where t_0 is an appropriate time interval.

On the other hand, if we let v_m be the motor's velocity, and F be the thrust, then it follows from our assumption of no losses that

$$P_{IN} = v_m F. \tag{5.130}$$

We first expand du/dt into the following form:

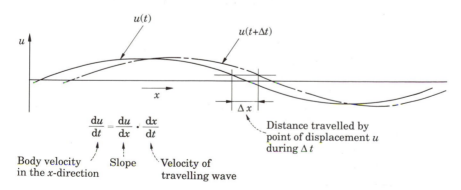

FIG. 5.32. Relation between differentials.

$$\frac{du}{dt} = \frac{du}{dx} \cdot \frac{dx}{dt} \qquad (5.131)$$

where dx/dt is the travelling wave's velocity (see Fig. 5.32), which is equal to $v\lambda$. If we let $dS = dx\,dy$, then

$$P_{IN} = \int_{S} \left(\frac{1}{t_0} \int_{t}^{t+t_0} f \frac{du}{dt}\, dt \right) dS$$

$$= v\lambda \int_{x} \int_{y} \frac{1}{t_0} \int_{t}^{t+t_0} f \frac{du}{dx}\, dt\, dx\, dy. \qquad (5.132)$$

Assuming uniform conditions in the y-direction (i.e. widthwise), with the beam width denoted by D,

$$P_{IN} = v\lambda D \frac{1}{t_0} \int_{t}^{t+t_0} \int_{u(x)} f\, du\, dt. \qquad (5.133)$$

Here $\int_{u(x)} f\, du$ is the Stieltjes integral with parameter $u(x)$ over the entire x-domain (this is discussed in Chapter 8). If we let t_0 converge to zero, and assume that there are m wave cycles in the integral's domain, and furthermore, let $\oint_{x} f\, du$ denote the integral over one cycle, then

$$P_{IN} = mv\lambda D \oint_{x} f\, du. \qquad (5.134)$$

We can equate this now with eqn (5.130) to obtain

$$P_{IN} = mv\lambda D \oint_{x} f\, du = v_m F. \qquad (5.135)$$

So the thrust F can be expressed as

$$F = \frac{mv\lambda D}{v_m} \oint_{x} f\, du \quad \left(= \frac{mv\lambda D}{v_m} \int_{0}^{\lambda} f \frac{du}{dx}\, dx \right). \qquad (5.136)$$

The thrust, however, can be considered as the integral of the tangential force over the surface (see eqn (8.79.) That is,

$$F = D \int_{x} f\, dx. \qquad (5.137)$$

Equations (5.136) and (5.137) can be equated, assuming there are no losses due to friction. Thus

$$\frac{v\lambda}{v_m} \int_{0}^{\lambda} f \frac{du}{dx}\, dx = \int_{0}^{\lambda} f\, dx \qquad (5.138)$$

for one wave cycle ($m = 1$).

If we assume here that u can be expressed as

$$u = -u_0 \sin(\omega t - kx) \qquad (5.139)$$

then we have

$$\frac{du}{dx} = ku_0 \cos(\omega t - kx)$$

$$= \frac{2\pi}{\lambda} u_0 \cos(\omega t - kx). \qquad (5.140)$$

Substituting this into the left-hand side of eqn (5.138), we get

$$\frac{2\pi \nu u_0}{v_m} \int_0^\lambda f \cos(\omega t - kx) \, dx = \int_0^\lambda f \, dx. \qquad (5.141)$$

This equation can be considered valid under certain restricted conditions, though it does not necessarily hold in all cases. Let us discuss three cases.

(1) *For a stator and a rotor coming into contact at wave crests as shown in Fig. 5.33.* In this case the motor velocity v_m equals the maximum tangential velocity at the wave's crest, which is given by

$$v_m = \left(\frac{du}{dt}\right)_{max} = \omega u_0 = 2\pi \nu u_0. \qquad (5.142)$$

Let us next investigate eqn (5.141). Near the wave's crest, du/dx has a maximum value. In other words, $\cos(\omega t - kx) \simeq 1$ (see eqn (5.140)). Outside this area, the tangential force f (applied by the stator to the rotor) is zero, because there is no contact. In this case the integration portion of eqn (5.141) becomes the same as the right-hand side, and we get

$$2\pi \nu u_0 / v_m = 1. \qquad (5.143)$$

This is equivalent to eqn (5.142); eqn (5.139) holds.

(2) *For a rotor made from an elastic material (e.g. rubber) as another sample that makes eqn (5.142) hold.* As shown in Fig. 5.34, contact occurs when $u > 0$ and no sliding is assumed to take place. The stator is deformed to the left ($\leftarrow$) at the crest, and to the right ($\rightarrow$) around $u = 0$. The velocity of the rotor or slider in this case will be less than $2\pi \nu u_0$. Let the displacement and tangential force be given by

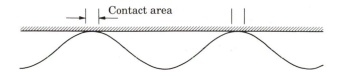

FIG. 5.33. Losses are minimal when contact occurs only near wave crests.

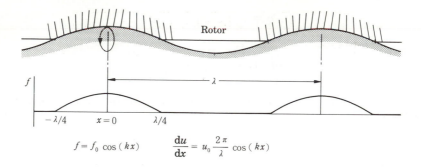

$$f = f_0 \cos(kx) \qquad \frac{du}{dx} = u_0 \frac{2\pi}{\lambda} \cos(kx)$$

Fig. 5.34. A rotor made from flexible rubber moves without sliding.

$$u = u_0 \sin(kx) \tag{5.144}$$

and

$$f = f_0 \cos(kx) \tag{5.145}$$

respectively. Taking account of $k = 2\pi/\lambda$, the thrust per wavelength (or for $m = 1$) given by eqn (5.136) derived for loss-free cases is expressed as

$$F = \frac{2\pi u_0 \nu f_0 D}{v_m} \int_{-\lambda/4}^{\lambda/4} \cos^2(kx)\, dx$$

$$= \frac{\pi \nu u_0 f_0 \lambda D}{2 v_m}\,. \tag{5.146}$$

On the other hand, the thrust can be given as the integral of surface forces (eqn 5.137), which is

$$F = f_0 D \int_{-\lambda/4}^{\lambda/4} \cos kx\, dx = \frac{\lambda}{\pi} f_0 D. \tag{5.147}$$

Equating these two expressions, we obtain

$$\frac{\pi \nu u_0 f_0 \lambda D}{2 v_m} = \frac{\lambda f_0}{\pi} D \tag{5.148}$$

$$v_m = \frac{\pi^2}{2} \nu u_0 = \left(\frac{\pi}{4}\right) 2\pi \nu u_0. \tag{5.149}$$

In other words, with no losses taking place, the rotor's velocity is the maximum value $2\pi \nu u_0$ multiplied by $\pi/4$.

(3) *For the case where sliding is prevented yet losses occur.* This is a case where eqn (5.141) does not hold. Even if the rotor and stator contact each other only at the wave's crests, and sliding is prevented from taking place,

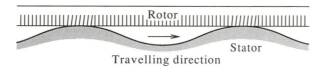

Fig. 5.35. Lower velocity due to shear strain, despite rotor and stator contact only around the wave crests.

the rotor's velocity can be lower than $2\pi\nu u_0$ in some cases. This can happen if the rotor is made of, say, a polymer and deformations arise due to shear forces.

As shown in Fig. 5.35, the tangential force due to the stator's elliptical motion creates shear strains in the rotor's contact area. The strain exists only during contact. Because of these local strains, the rotor is prevented from reaching the theoretical maximum velocity. In such cases, losses occur in the contact area, even though there is no sliding. We shall assume that the stator's tangential force is transmitted directly to become the frictional force experienced by the rotor, and that this appears as the torque on the rotor's shaft. In other words, the torque T generated by the stator is assumed to be equal to the output torque.

If we let u_s be the stator's surface velocity in the tangential direction, and R be the effective radius of the contact surface, then the work done by the stator is $2\pi R u_s T$. Similarly, if we let u_r be the rotor's velocity at the outer perimeter on the non-contact side, the output is $2\pi R u_s T$. Since $u_r < u_s$, a portion of the work done by the stator is lost, with the remainder becoming the output. In other words, some of the power is lost somewhere in the process. A likely cause of this loss is that, as the rotor's contact area experiences oscillatory strains of high frequencies, the strain energy is steadily transformed into heat because of the rotor's viscoelastic nature. Moreover, contrary to the assumption that we have made, the output torque is likely to be lower than the torque generated by the stator. That is, part of the frictional force experienced by the rotor is used to deform and accelerate the rotor's contact area. The strain energy thus created is dissipated as heat when the contact is broken and the strain removed.

There are also other factors that cause losses. If the rotor is compressed (vertically) at the contact areas, the strain energy thus created is also dissipated as heat when that part of the rotor separates from the stator surface and rebounds to its initial shape.

Thus we have seen that losses due to sliding and deformation occur around the contact surface between the rotor and stator. There are several approaches to analyse the phenomena of losses quantitatively. One way may be to examine the behaviour of the contact surface in close detail. We shall conclude our discussion for now, however, and treat the subject using equivalent circuits in the next chapter. Later, in Chapter 8, we shall continue theoretical discussion on losses and examine their causes.

Reference

1. Timoshenko, S. (1955) *Strength of materials*. Part II, *Strength theory and problems*, (3rd edn). Van Nostrand, New York.

6. Equivalent-circuit analysis for the travelling-wave motor

The basic principles of the ultrasonic motor, which were discussed in Chapter 1, are not difficult to grasp, at least qualitatively. They are much simpler than the principles of a.c. electromagnetic motors such as the induction motor or the hysteresis synchronous motor. (This is the subject of Chapter 8.) A mechanical engineer, for instance, should have no difficulty in understanding how the ultrasonic motor works.

However, as we have seen in Chapters 2 to 5, the details involved are fairly complex and many aspects of the motor are as yet unresolved. In this chapter we present a method that allows us to determine the characteristics of a travelling-wave motor in a quantitative manner, while leaving room for a few unresolved problems. For this purpose, we shall advance the theory of equivalent circuits discussed in Chapters 2 and 3.

In Chapter 4 we presented an equivalent circuit for the wedge-type motor which involves some non-linear elements. The characteristics can be determined by solving a set of non-linear differential equations that represent the circuit. Instead of demonstrating this solution, however, which would occupy too much space and diverge from the purpose of this text, we limit our discussion to the travelling-wave motor, which is considered a more practical type.

The equivalent circuit developed for the ultrasonic motor is based on a somewhat different concept from those for electromagnetic motors. In this connection, Section 8.6 will be devoted to a comparison with a d.c. motor, to derive several distinct differences between the new and conventional motor.

6.1 Utility of equivalent circuits

Of the various types of a.c. motor, the squirrel-cage induction motor is the most widely used and extensively studied. A number of people were involved in the invention, around 1885, of the induction motor. The greatest contribution was made by Nikola Tesla, who was born in Croatia and later became a US citizen. He invented what are known as single-phase and two-phase induction motors.

The three-phase squirrel-cage induction motor was invented by a German scientist, Dolivo von Dobrovolski, in 1889. This type of motor in particular owes much of its subsequent development to the use of equivalent circuits for analysis of its characteristics. Equivalent-circuit methods originated not from the original inventors of motors but from later scholars and engineers who studied them. These methods have proved to be very useful tools and are still widely used today for solving complex problems.

Although the authors have extensively discussed the subject of equivalent circuits for ultrasonic motors and analysed several cases using desktop computers, there still remain some unresolved problems. In the following sections we construct an equivalent circuit for the travelling-wave motor step by step, analyse it, and then compare the results with laboratory measurements. We present the equivalent circuit not as the final solution but rather to demonstrate its value as an analytical tool.

6.2   Equivalent circuit for the travelling-wave motor

We begin our discussion with an equivalent circuit for a stator carrying no rotor.

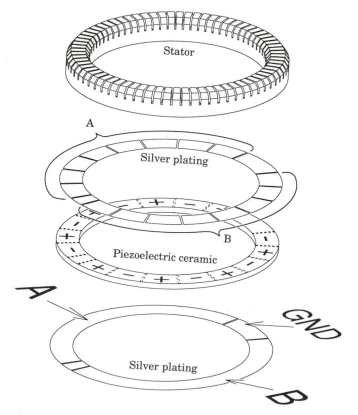

FIG. 6.1.   Stator construction of a two-phase motor.

6.2.1 Stator without the rotor

The travelling-wave motor normally has a two-phase construction as shown in Fig. 6.1. Although a three-phase scheme has been attempted in the laboratory, it has not yet shown significant merit. Each phase, A and B, comprises forward-poled sections (denoted by $+$) and reverse-poled sections (denoted by $-$). (Refer to Section 3.1.3 for the definition of 'poling'.) Figure 6.2(a) shows a stator detached from the rotor, with its equivalent circuit in (b). Each phase has essentially the same circuit that was presented earlier in Fig. 3.20.

Figure 6.3 shows the standing wave seen at two phases with only phase

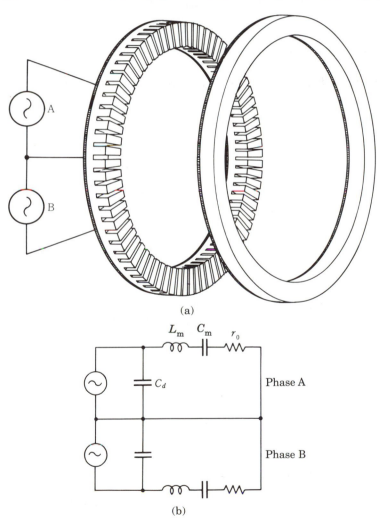

(a)

(b)

FIG. 6.2. (a) Ultrasonic motor with rotor detached; (b) equivalent circuit.

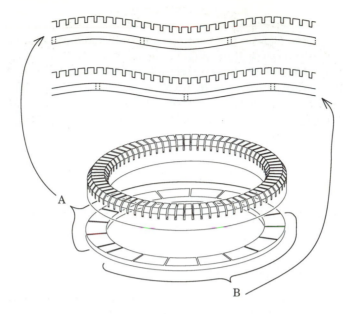

FIG. 6.3. When only phase A is excited.

A excited at the moment in the polarity shown. The forward-poled segments expand, and the reverse-poled ones contract. If the excitation polarity is reversed, the forward-poled sections will contract and the reverse sections expand. Therefore when phase A is excited by an alternating voltage, a standing wave is created. For the phase-B ceramic element, the midpoint of each segment corresponds to a node of the standing wave excited by phase A. For each segment, one-half is forced to expand while the other half is made to contract, so the total length remains the same and no electricity is generated. In other words, no mutual electric effect takes place between phases A and B. Thus the two phases are independent of each other in this arrangement.

The elements in this equivalent circuit have the following meanings:

C_d: blocking capacitor — acts as a regular dielectric

L_m: equivalent inductor — represents the mass effect of the ceramic body and ring

C_m: equivalent capacitance — represents the spring effect of the ceramic body and ring

r_0:   resistance representing losses that occur within the ceramic body and ring.

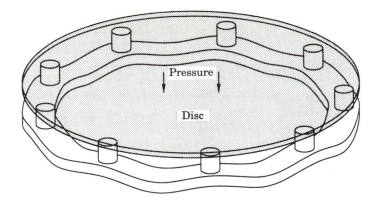

FIG. 6.4. Supports positioned at every other phase-A loop and pressure applied with disc.

6.2.2 With a connected load

Consider a load as shown in Fig. 6.4, in which supports are positioned along the phase-A stator ring at every other vibrational loop and a disc is pressed against the entire assembly. The equivalent circuit for this condition is shown in Fig. 6.5. The elements newly added to this circuit are;

C_K: equivalent capacitance representing the spring effect of the support — disc assembly

E_f: equivalent voltage representing the applied pressure on the assembly

L_K: equivalent inductance representing the mass of the assembly.

The newly added section can be interpreted as follows. The motion created in the stator ring causes deformation or movement in the disc assembly. To be precise, it is the difference between the applied pressure and the force associated with the stator motion that determines the disc assembly's motion. Stated in terms of the equivalent circuit this translates as: the difference between the voltage across the condenser C_K and the d.c. voltage E_f is applied to L_K and this determines the current (or velocity in mechanical terms).

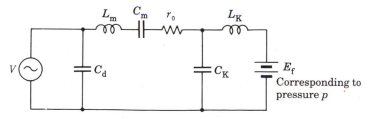

FIG. 6.5. Equivalent circuit for the configuration of Fig. 6.4.

Since electrical (e.g. voltage or current) and mechanical (e.g. force or displacement) quantities can both be expressed in the equivalent circuit, it is ideally suited for analysing an actuator that contains piezoelectric ceramic elements. (As we have seen, mechanical parameters are converted into equivalent electrical ones.) Table 6.1 shows the one-to-one correlations between mechanical and electrical parameters. In subsequent discussions, we shall mainly use the electrical quantities and occasionally refer to their mechanical equivalents when necessary.

6.2.3   Treatment of d.c. voltage

In the circuit of Fig. 6.5, the d.c. voltage source E_f, which represents the pressure applied to the stator as in Fig. 6.4, does not generate a steady current. This means that, mechanically, the pressure acting on the disc does not create a sustained unidirectional motion. However, this pressure causes transitory motions in the supports, pressure disc or stator ring, in short spurts. In many cases it is permissible to omit such d.c. voltage sources (and the charge they create) from the equivalent circuit. This would then result in the circuit in Fig. 6.6.

6.2.4   Complete motor

We now add the stator. As shown in Fig. 6.7, the rotor is driven not by the vertical vibrations of the stator but by sideways movements of the contact surface. To examine this phenomenon in more detail, consider the role played by a single comb-tooth, shown in Fig. 6.8. This model is based on the interpretation shown in Fig. 5.29.

The mechanism involved can be broken down into two parts: (1) the comb-tooth makes contact with and separates from the rotor by vertical motion; (2) during contact, the vertical motion is converted to motion in the horizontal direction by a lever principle, and this drives the rotor. Stated in terms of the equivalent circuit: when the comb-tooth is not in contact, it moves unrestrained and corresponds, in Fig. 6.8(b), to the switch connected to the 'separation' side. When in contact, the comb-tooth's vertical

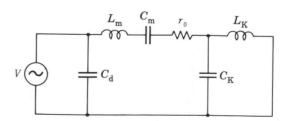

FIG. 6.6.   Equivalent circuit in which d.c. voltage source E_f is removed.

Table 6.1. Correspondence between electrical and mechanical quantities

Electrical		Mechanical			
		Linear		Rotary	
Quantity and symbol	Units	Quantity and symbol	Units	Quantity and symbol	Units
Voltage V, E	V	Force f, F	N	Torque T	N m
Current I	A	Velocity v	m s^{-1}	Angular velocity ω	rad s^{-1}
Charge Q, q	C	Distance x	m	Angle Θ	rad
Inductance L	H	Mass M, m	kg	Inertia J	kg m^2
Resistance R, r	Ω	Viscosity D	N s m^{-1}	Viscosity D	N m s rad^{-1}
Capacitance C	F	Spring constant K	N m^{-1}	Angular stiffness	N m rad^{-1}
Force factor A	C m^{-1}	Force factor A	N V^{-1}	Torque factor	N m V^{-1}

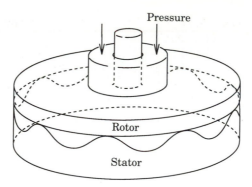

Fig. 6.7. When a rotor is pressed against the stator, a torque is applied to the rotor.

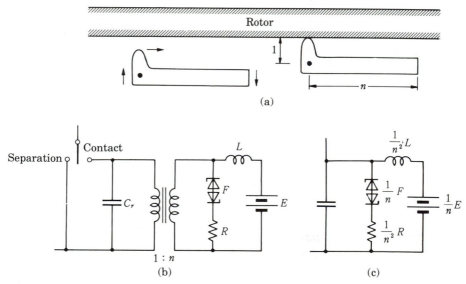

Fig. 6.8. Equivalent circuits of lever principle at contact surface: (a) model of a comb-tooth lever; (b) equivalent circuit 1; (c) equivalent circuit 2.

motion is transmitted to the rotor by the lever's 'transformer effect'.

The condenser C_r performs the same role as C_K in Fig. 6.5. In an ideal situation, the comb-tooth's tangential motion will be transmitted to the rotor without any losses; in actuality, some sliding takes place and heat is generated as a result of frictional losses. This effect is represented by the two diodes and resistor connected in parallel with the transformer. E is the torque load on the comb tooth. The inductor between E and the 'friction' elements represents the rotor's inertia (or moment of inertia). If we remove the transformer and adjust the elements to the appropriate equivalent parameters, we have Fig. 6.8(c).

Next we need to consider the two phases, A and B, which together drive the rotor. Since the two phases do not simply 'take turns' to run the motor, the equivalent circuit in Fig. 6.8(c) should correspond to a state when both phases are associated with the rotor–stator contact, with the parameters adjusted accordingly. Moreover, the equivalent circuit should include a rectifier to represent the unidirectional motion of the rotor. However, this cannot be achieved by simple semiconductor diode circuits that can perform mere half-wave or full-wave rectification. In the ultrasonic motor, subtle rectification is effected by mechanical means. This is shown by the rectifier circuit in Fig. 6.9.

L_r, which represents the rotor's inertia, is necessary for analysing motion if the ultrasonic motor is used as a servomotor. Under steady-state running conditions, however, only direct currents will flow through this element and there will be no voltage drop (i.e. no torque required for acceleration or deceleration), so we can eliminate it from the circuit.

D_r and R_r are elements representing frictional and viscous losses in the bearing and other related parts. Whereas D_F and R_F, elements in parallel

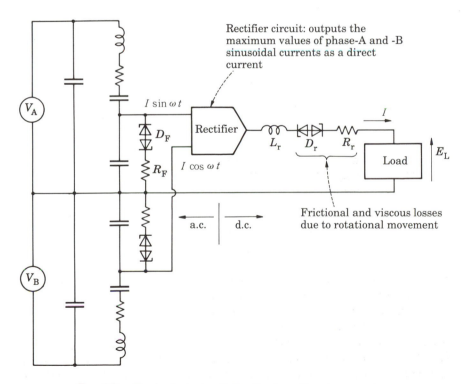

FIG. 6.9. Equivalent circuit for the travelling-wave motor.

with C_r, represent losses due to sliding between the stator and rotor, D_r and R_r represent mechanical losses which result from the rotor's movement. Although it is fairly easy to understand the physical significance of the elements and their relationships in the equivalent circuit in Fig. 6.9, its computational analysis is laborious. The circuit contains a non-linear element D_F, which would require solving a set of three second-order differential equations.

The authors have analysed similar problems using a mainframe as well as a desktop computer. The process was tedious, to say the least. In the next section, therefore, we introduce a few approximations to construct a 'simplified equivalent circuit' and keep computations at a manageable level.

6.3 Simplified equivalent circuit

For each phase, A and B, losses and outputs are quantities that vary sinusoidally. Although the phase difference in current or voltage between the two phases is 90°, it is 180° for losses or output power. This is because the latter quantities are expressed in terms of the squares of either the current or the voltage (see Fig. 6.10). If losses or output for both phases are

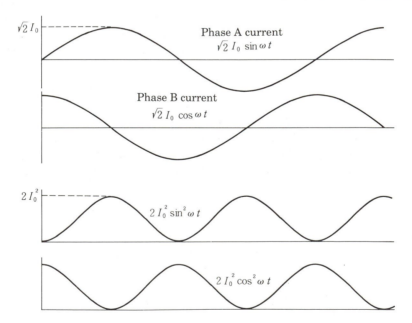

FIG. 6.10. Phase-A and -B currents have a 90° phase difference: the square of these currents will consist of a d.c. component and an a.c. component, with a phase difference of 180°. If the two phases are added together, a constant value is obtained.

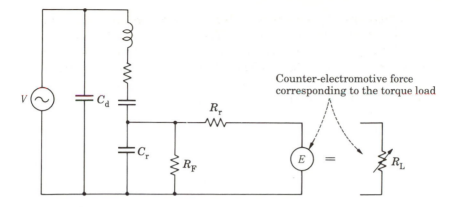

FIG. 6.11. Equivalent circuit for a single phase.

added together, the resultant is a constant. This is the direct current flowing in the d.c. portion of the circuit in Fig. 6.9.

Now we separate the equivalent circuit for phases A and B. We then remove the diodes and let resistors represent all of the mechanical losses so that we can treat the circuit as a simple a.c. circuit. The resulting circuit is shown in Fig. 6.11. The rectifier element in Fig. 6.9 outputs a d.c. current I when the input currents from phases A and B are $I \sin \omega t$ and $I \cos \omega t$ respectively. However, we shall treat the equivalent circuit in Fig. 6.11 as a normal a.c. circuit. This a.c. equivalent circuit can be further simplified by the following two steps:

(1) *Removal of C_r.* In Fig. 6.11, C_r is the equivalent capacitor represen-ting the stator's strain created in response to the reaction by the load torque applied to the rotor. We can eliminate this capacitor if its effect is small relative to the parallel equivalent resistor. This will result in Fig. 6.12. This capacitor can affect the resonant frequency of the system. Although the resonant frequency is higher with applied rotor pressure than without it, the difference amounts to only ~5 per cent.

(2) *Use of a reactor for power factor improvement.* The blocking capacitance C_d lowers the power factor. It would have no effect on the

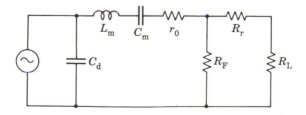

FIG. 6.12. Simplified equivalent circuit.

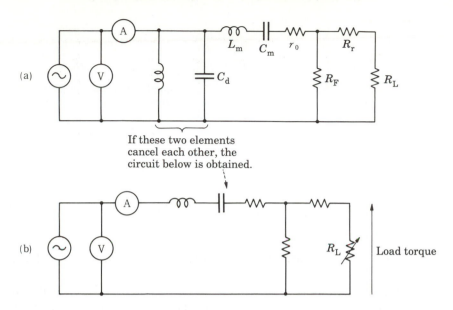

(a)

(b)

If these two elements
cancel each other, the
circuit below is obtained.

Load torque

FIG. 6.13. Voltmeter (V) and ammeter (A) for laboratory measurements: terminal
voltage on R_L corresponds to torque load.

motor's efficiency if there were no line resistance connected between the
power supply and the motor. However, under actual conditions, a low
power factor adversely affects the efficiency, owing to the power source's
high internal resistance or resistance in the motor's control circuit. To
improve the power factor, we can place a reactor in parallel with the blocking capacitance as shown in Fig. 6.13(a). The voltmeter and ammeter in this
circuit are shown for measurement purposes. If the blocking capacitor's
effect is completely cancelled by the reactor, we can use the equivalent
circuit in (b).

6.4 Treatment of applied pressure and R_F

In Figs 6.11–6.13, the element R_F represents friction and the amount of
sliding between the rotor and stator. Friction in turn is related to the applied
pressure (as discussed in Chapter 2). Although friction can be represented
by eqn (2.1) (a) or (b) for a simple case, we need a model for assessing friction more accurately for the ultrasonic motor. In the following, we present
such a model, which should be viewed as a proposal rather than a final
solution.

In our case, R_F is a function of the load torque and applied pressure.
Generally speaking, a higher load torque results in a lower R_F, i.e. more

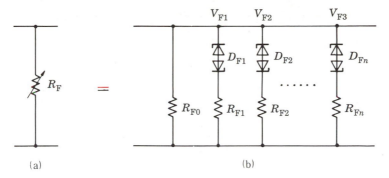

FIG. 6.14. (a) R_F, a function of load and pressure; (b) its model. V_{F1}, V_{F2} ..., V_{Fn} are the breakdown voltages for D_{F1}, D_{F2}, ..., D_{Fn}, respectively.

sliding. This characteristic can be achieved using a model such as shown in Fig. 6.14.

R_{F0} represents sliding which takes places even at a light load. An increase in the load is represented in the equivalent circuit by an increase in the terminal voltage across the R_F's. If this voltage exceeds V_{F1}, diode D_{F1} breaks down and current will flow through R_{F1}. If the torque load is increased further and the terminal voltage of R_F exceeds V_{F2}, diode D_{F2} breaks down. Thus as the load torque is gradually increased, the diodes will break down one by one, and the overall resistance will be lowered.

The more diode–resistor elements are placed in parallel, the more accurate the model becomes, although there is a certain limit above which further additions become pointless. The authors have not yet established values for the breakdown voltages V_{F1} to V_{Fn} and the resistances R_{F0} to R_{Fn}. However, we can make the general observation that if the applied pressure is low, the maximum torque will also be low. So low R_{Fm} values should be used for the lower voltages of V_{Fm}. In actual motors, however, there is an appropriate range for the drive voltage for a given pressure, below which the motor may not function properly.

6.5 Torque coefficient and problems related to the force factor

As we have already noted, the current and terminal voltage of R_L are proportional to the rotor's speed and torque output respectively. We shall examine these relationships more closely in this section.

6.5.1 Definition of torque factor α_T

First we define the torque factor α_T to express the relation between a single phase's output torque $T/2$ and voltage E_L:

$$T/2 = \alpha_T E_L. \tag{6.1}$$

Then the relation between current I_L and the rotor's angular velocity N can be shown to be

$$I_L = \alpha_T N \tag{6.2}$$

which follows from the fundamental relation

$$\underset{\text{(mechanical output)}}{TN} \quad = \quad \underset{\text{(electrical output)}}{2E_L I_L}. \tag{6.3}$$

6.5.2   Torque factor α_T versus force factor A

In section 3.3 we showed the equivalent circuit for the piezoelectric ceramic body (using a four-terminal network), in which force factor A could be interpreted as either the force created by unit applied voltage (1 V) or the current created by unit velocity (1 m s^{-1}).

Using the former interpretation, the force at the rotor's perimeter is $A_0 E_L$ if a voltage E_L is applied to the equivalent circuit's resistance R_L (using A_0 for the force factor). So with a rotor radius of R (using the stator ring's mean radius as the effective radius), torque T is given by

$$T = 2RAE_L. \tag{6.4}$$

Therefore, by comparing with eqn (6.1), the torque factor α_T is given by

$$\alpha_T = RA. \tag{6.5}$$

On the other hand, using the second interpretation of A if v is the rotor's peripheral velocity when current I_L flows through R_L, then

$$I_L = Av. \tag{6.6}$$

If we let N be the angular velocity, then $v = NR$, so

$$I_L = ARN. \tag{6.7}$$

By comparing the above equation with eqn (6.2), we can check eqn (6.5). As stated in Section 3.3, the force factor depends not only on the shape, dimensions and other factors of the piezoelectric ceramic piece, but also on the type of metal with which it is combined. Here we shall define the force factor A as follows: the velocity $(\mathrm{d}u/\mathrm{d}t)$ of the stator's wave crest is $1/A$ (m s^{-1}) when the effective current for phase A is 1 A, assuming that both phases A and B are running in balance (i.e. the same condition but for a 90° difference).

6.6 Sample measurements and calculations of characteristics

In this section we compare some measurements of characteristics with theoretical values to see whether they confirm the theory presented so far.

6.6.1 The effect of driving frequency

Figure 6.15 shows the relation between velocity and torque measurements when the driving frequency was varied within a range slightly above the resonant frequency. Phosphor bronze and NEPEC-61 piezoelectric ceramic (Tokin Corporation) were used for the stator's elastic body and piezoelectric element respectively. The stator had an outer diameter of 60 mm, nine waves on the periphery, and 72 teeth. For the rotor, an aluminium body coated with of engineering polymer was used. Measured parameters are given in the caption.

A graph of current, phase angle, input, output, and efficiency for a driving frequency of 40.33 kHz is shown in Fig. 6.16. The maximum efficiency is low, ~16 per cent. This is caused by large profile irregularities and highlights the importance of flatness and surface smoothness of the stator and rotor. Since the test model was made so that it could easily be dismantled for adjusting the applied pressure, it is possible that repeated dismantling and assembly caused the large surface irregularities.

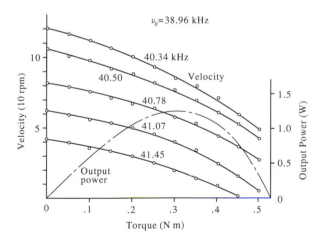

Fig. 6.15. Velocity–torque characteristics at various driving frequencies (example). Applied voltage, 100 V (r.m.s.); pressure, 132 N; resonant frequency ν_0, 38.96 kHz; L_m, 163 mH; C_m, 102 pF; C_d, 8.56 nF (values of the last four were measured on an independent stator). Output power is plotted for 41.07 kHz.

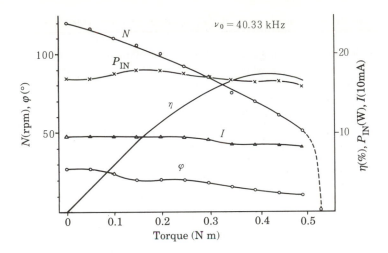

Fig. 6.16. Velocity (N), input (P_{IN}), current (I), phase angle (φ), and efficiency (η); a compensating reactance of 1.77 mH is connected to each phase.

6.6.2 Sample calculations of characteristics

The characteristic curves in Figs 6.17 and 6.18 were derived from calculations using the model for R_F proposed in Section 6.4. Figure 6.17 shows the velocity–torque characteristics when the driving frequency was varied from 40.33 to 41.25 kHz, while Fig. 6.18 shows the calculated efficiency and current curves at 40.33 kHz. For R_{Fn}, which is assumed to decrease as n increases, two equations are considered: $R_{Fn} = 15\,000/(1 + 0.6n)$ Ω and $R_{Fn} = 20\,000/(1 + n)$ Ω, for which we used force factors $A = 0.23$ C m^{-1} and $A = 0.24$ C m^{-1} respectively.

It should be noted that while we used $L_m = 163$ mH and $C_m = 102$ pF as the measured values for the stator alone, lower capacitance values, 97.0 and 96.5 pF, were used to determine characteristics when the rotor is attached and pressure applied. With these capacitances, the theoretical results approximate the experimental.

6.6.3 A problem associated with current

Figure 6.19 shows characteristic curves for a recent laboratory-made motor which has attained a maximum efficiency as high as 50 per cent by careful selection and machining of the rotor material. The curves are taken with a constant applied voltage. There is one puzzling feature in these graphs, however: the current increases as the torque is raised. Theoretically, in an ideal motor with a high R_F value, the current should decrease as the

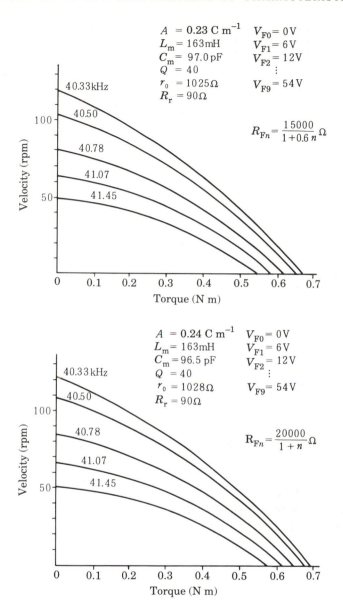

$$A = 0.23\ \text{C m}^{-1} \quad V_{F0} = 0\,\text{V}$$
$$L_m = 163\,\text{mH} \quad V_{F1} = 6\,\text{V}$$
$$C_m = 97.0\,\text{pF} \quad V_{F2} = 12\,\text{V}$$
$$Q = 40 \quad \vdots$$
$$r_0 = 1025\,\Omega \quad V_{F9} = 54\,\text{V}$$
$$R_r = 90\,\Omega$$

$$R_{Fn} = \frac{15000}{1+0.6\,n}\ \Omega$$

$$A = 0.24\ \text{C m}^{-1} \quad V_{F0} = 0\,\text{V}$$
$$L_m = 163\,\text{mH} \quad V_{F1} = 6\,\text{V}$$
$$C_m = 96.5\,\text{pF} \quad V_{F2} = 12\,\text{V}$$
$$Q = 40 \quad \vdots$$
$$r_0 = 1028\,\Omega \quad V_{F9} = 54\,\text{V}$$
$$R_r = 90\,\Omega$$

$$R_{Fn} = \frac{20000}{1+n}\ \Omega$$

Fig. 6.17. Sample calculations of velocity–torque characteristics.

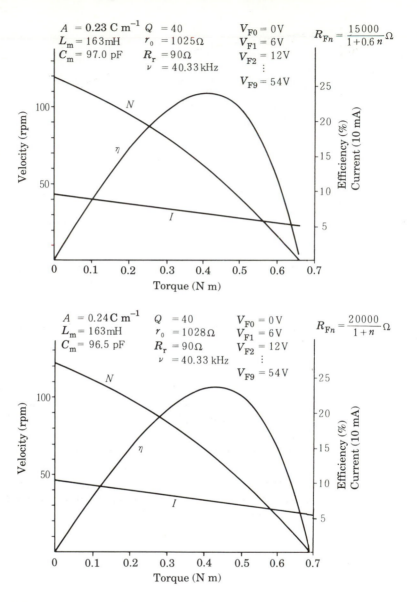

Fig. 6.18. Sample calculations of velocity, efficiency and current as functions of torque.

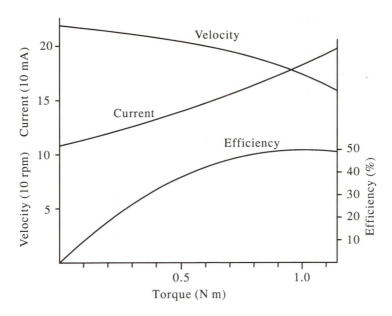

FIG. 6.19. Characteristics of a recent test motor. The current increases with load torque.

torque rises. This phenomenon of increasing current with increasing torque cannot be explained even if we take into consideration the incompleteness of the compensation reactor, which is placed so as to cancel the leading current due to the blocking capacitor.

Even if we had a model in which the sliding resistance decreased as the torque increased, the current would still decrease. One possible interpretation is that the equivalent capacitance C_m varies with torque. This suggests that there are imperfections in the equivalent-circuit theory discussed in this chapter. Indeed, the theory is still under development.

7. Design, assembly, and testing of a prototype ultrasonic motor

We have so far discussed the theoretical aspect of the ultrasonic motor. In this chapter, our subject is the actual assembly and testing of a prototype ultrasonic motor. In Section 3.9 we examined various limiting conditions as factors affecting motor design in general. Here we show some calculation examples and make comparisons with theory based more or less on a complete motor.

Design experience with ultrasonic motors, unlike that with electro-magnetic motors, has been very limited and there still remain many unre-solved questions about its characteristics. Thus the authors feel that it is unrealistic to present a design theory for the ultrasonic motor in a final and orderly manner, and it is their wish that the material presented in this chapter may serve as a starting point for further research.

A testing system designed for use on mass-production lines is also described here.

7.1 Exploded view of the ultrasonic motor

It is important to have some knowledge of hardware to understand motor design, and for this purpose an exploded view of the ultrasonic motor is shown in Fig. 7.1. This was designed for use for the roll curtain drive without reduction gears. Figure 7.2 shows photographs of parts and an assembled motor. A mechanical engineer should be able to visualize the motor assembly from these figures. Processing and electrical treatment of the piezoelectric ceramic components are shown in Figs 7.3 and 7.4. Photo-graphs (1)–(9) in Fig. 7.4 depict the processes indicated in the flow chart of Fig. 7.3.

7.2 Basic design considerations

As we have seen in the previous chapter, the maximum efficiency of the travelling-wave motor is not very high. Despite efforts to improve the efficiency, it currently remains around 50% in laboratories. If dramatic improvements were to be made in the future, the maximum efficiency should occur at low speeds and high torque. (This will be discussed in the next chapter.) However, with current motors, maximum efficiencies are

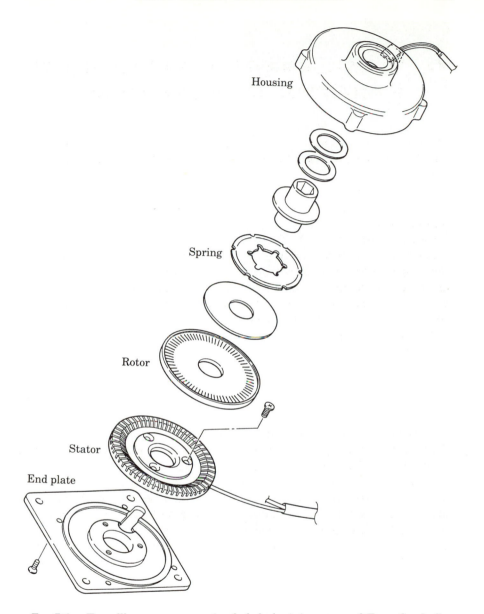

Housing

Spring

Rotor

Stator

End plate

Fig. 7.1. Travelling-wave motor (exploded view) (courtesy of Toso Co. Ltd).

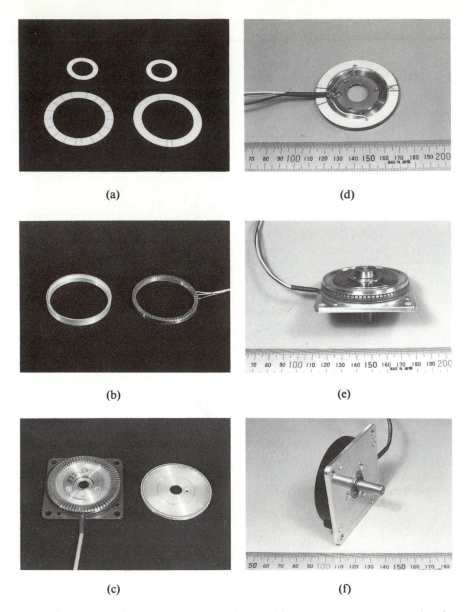

Fig. 7.2. Parts and completed units of the travelling-wave motor: (a) piezoelectric ceramic ring with inner hole cut out and electrodes installed (view of both sides); (b) rotor ring (left) and stator ring (right); (c) stator and rotor; (d) wiring connected; (e) assembled motor; (f) assembled motor with housing.

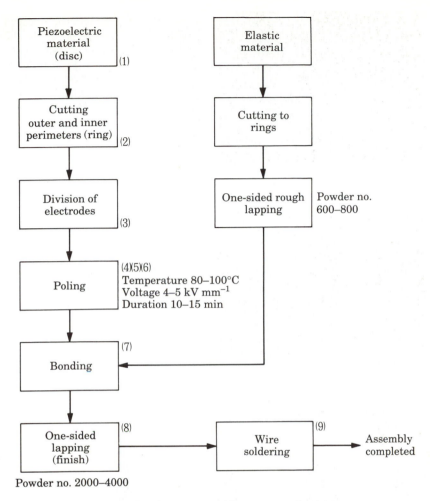

FIG. 7.3. Stator assembly process scheme.

obtained at speeds of about half of the no-load speed. With these pre-liminaries, we shall examine some basic factors affecting design dimensions of the travelling-wave motor.

(1) *Diameter and output*. Basically, as shown in the curve in Fig. 7.5(a), the maximum output obtained under normal voltages is proportional to the square of the motor diameter:

$$\text{maximum output} \propto (\text{diameter})^2. \tag{7.1}$$

Maximum outputs are obtained at speeds of half no-load speed, as noted in Fig. 6.16.

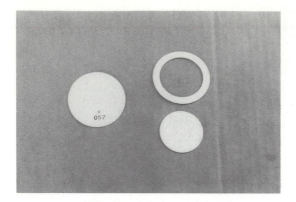

(1) Piezoelectric material. The piezoelectric plate has silver electrodes baked on to both sides (thickness ~0.01 mm). Some manufacturers provide them already polarized, in which case the poles are indicated. The piezoelectric plate is shown before (left) and after (right) inner hole has been cut out.

(2) Hole cutter for ceramic plate. The piezoelectric plate is installed in the retainer and a GC no. 600 powder-kerosene mixture is applied on to its surface before cutting. Cutting requires 30 s to 1 min.

(3) Electrode divider. The piezoelectric ring is set on a table and the silver electrode coating is removed with an abrasive cutting wheel. Maximum diameter for ring: $\phi80$. Number of divisions: 18, 16, 14, 12, 10, 8.

FIG. 7.4.   (a) Stator assembly process: steps 1–3.

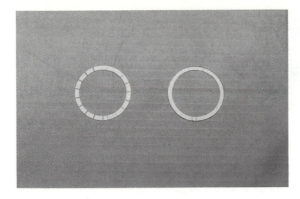

(4) Electrodes arranged on piezoelectric ring. Left: electrodes for polarization; this side will be bonded to stator. Right: the other side, to which wires are soldered.

(5) Polarization device. This polarizes all the electrodes on a piezoelectric ring in one step.

(6) Polarization. The polarization device and piezoelectric ring are immersed in insulating oil (80–100°C) and high voltage (~ 4–$5\,\text{kV}\,\text{mm}^{-1}$) is applied. Process requires 10–15 min.

FIG. 7.4.   (b) Stator assembly process: steps 4–6.

(7) Bonding jig. Bonded under high pressure at 80–90°C using Araldite, at room temperature using Loctite. Rings should be set in place quickly, especially when using Loctite No. 648, since it will begin setting in 10 s.

(8) One-sided lapping. Performed to achieve a high-precision surface finish (for flatness and surface roughness) on the elastic ring. Powder no. 2000–4000; applied pressure 1–2 N.

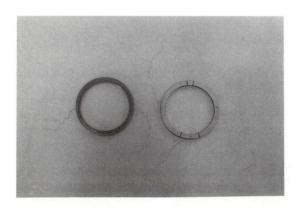

(9) Wiring soldered on. Wires should be as thin as possible (ϕ0.2) and attached with minimum amount of soldering compound to minimize their effect on the vibration. The finished stator is shown (front at left, back at right).

Fig. 7.4.   (c) Stator assembly process: steps 7–9.

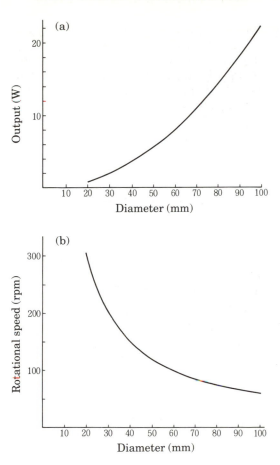

FIG. 7.5.   (a) Basic design consideration: relation between diameter and maximum output. (b) Speed is inversely proportional to diameter.

(2) *Torque and diameter.* The starting torque is proportional to the diameter cubed. If the rated torque (i.e. torque under normal running conditions) can be considered as equal to the starting torque multiplied by some coefficient, it follows that the rated torque is likewise proportional to the diameter cubed.

(3) *Speed and diameter.* The no-load speed is inversely proportional to the diameter. Since the rated speed is considered as the no-load speed multiplied by some coefficient, it can be considered to be inversely proportional to the diameter (see Fig. 7.5(b)).

7.3 Stator ring design

In the ultrasonic motor, the stator ring–piezoelectric ceramic ring assembly is made to resonate to create flexural travelling waves, which then drive the rotor. Flexural waves in beams were discussed in Chapter 5. In this section we determine dimensions for the stator ring and its parts, then examine the force factor A.

The theory we developed in Chapter 5 was based on the assumption of elastic bodies with a uniform density and Young's modulus. However, not only is the stator composed of two distinct materials (metal and piezoelectric ceramic), but the metal ring has a comb-tooth structure. This is pressed against an elastic load-carrying rotor. Moreover, some of the oscillation energy is lost through the stator's support structure. Since we do not have

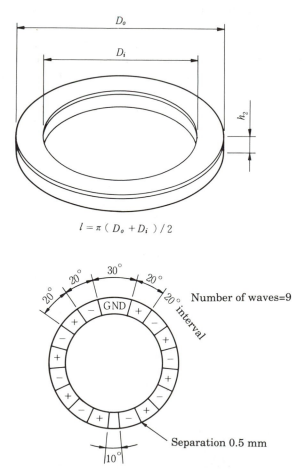

$$l = \pi \, (D_o + D_i) \, / \, 2$$

FIG. 7.6. Stator ring and electrode arrangement.

theories to analyse such complex structures, we must resort to simplifications and approximate theories to develop our argument.

7.3.1 Resonant frequency and dimensions

The frequency v_n for the nth mode wave for a beam of length l, which was given in eqn (5.91), is shown here again:

$$v_n = \frac{\pi n^2}{2l^2} \sqrt{\frac{EI}{\rho S}} \tag{7.2}$$

where E is the Young's modulus, ρ the density, I the second moment of area, and S the beam's cross-sectional area (we shall use S to avoid confusion with force factor A). Although this equation was originally derived for straight beams, it can also be applied to rings, for which the mean perimeter is substituted for l (see Fig. 7.6).

To determine v_n, we first derive an equation for the approximate value of the second moment of area. Assume a ring cross-section as shown in Fig. 7.7(a). Assume also that the ceramic and metal rings are firmly bonded together and vibrate as one unit. Since ceramics and metal have different Young's moduli, the cross-section can be transformed into an equivalent cross-section as shown in (b). Then b_2 and b_3 have the following relationship:

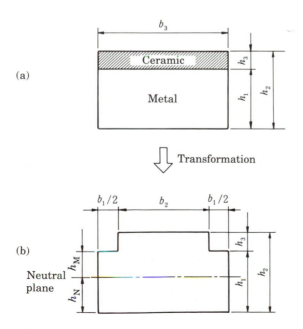

Fig. 7.7. Transformed cross-section for stator.

$$\frac{b_2}{b_3} = \frac{Y_{11}}{E_m} \tag{7.3}$$

where Y_{11} is the Young's modulus of the piezoelectric ceramic in the 1–1 direction (refer to Section 3.2) and E_m the Young's modulus of the metal.

The neutral plane's position h_N (see Fig. 7.7) is given by

$$h_N = \frac{b_1 h_1^2 + b_2 h_2^2}{2(b_1 h_1 + b_2 h_2)} \tag{7.4}$$

and the cross-sectional area S by

$$S = b_1 h_1 + b_2 h_2. \tag{7.5}$$

The second moment of area is given by

$$I = \frac{1}{3} b_2 (h_N^3 + h_M^3) + b_1 h_1 \left(h_M - \frac{h_1}{2} \right)^2 + \frac{b_1 h_1^3}{12} \tag{7.6}$$

where

$$h_M = h_2 - h_N. \tag{7.7}$$

If m wavelengths are generated on the ring's perimeter, the wavelength λ is given by

$$\lambda = \frac{\pi (D_o + D_i)}{2m} \tag{7.8}$$

where D_o and D_i are respectively the outer and inner diameters of the ring.

On the other hand, as discussed in Section 5.11, the oscillation mode n and the number of pole pairs m have the relation $n = 2m$, while

$$\lambda = 2l/n. \tag{7.9}$$

Hence eqn (7.2) can be expressed as

$$\nu = \frac{2\pi}{\lambda^2} \sqrt{\frac{EI}{\rho S}}. \tag{7.10}$$

Sample calculation of resonant frequency. A sample calculation was performed on a computer using the data given in Fig. 7.8. The BASIC program and its results are shown in Fig. 7.9. A resonant frequency of 54.52 kHz was obtained. For comparison, the same calculation was performed assuming that the Young's modulus for the ceramic and metal were equal, with 55.72 kHz as the result, which is not much different from the above value.

Measurements were then taken on a test model and a resonant frequency of 42.4 kHz was obtained — about 20% lower than the calculated result. This difference probably arises from the approximate nature of eqn (7.2). For a more exact design theory, the precise frequencies given by eqn (5.80)

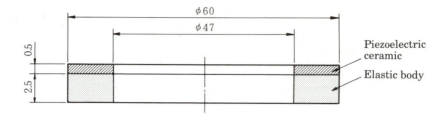

	Elastic body	Piezoelectric ceramic
Material	Phosphor bronze	NEPEC N-61
Density (kg m^{-3})	8.78×10^3	7.79×10^3
Young's modulus (N m^{-2})	11.2×10^{10}	7.6×10^{10}
Mechanical quality factor	3000	1800
Total mass (g)	34	

FIG. 7.8. Data for resonant frequency calculation.

```
100  REM COMPUTING RESONANT FREQUENCY
110  DO=60:  DI=47:  H1=2.5:  H3=.5:  H2=H1+H3    mm
120  Y11=7.6E+10:  EM=1.12E+11   :RHO=8.78  :  N=9
130  PI=3.1416
140  '**** COMPUTE ****
150  B3=(DO-DI)/2:  B2=(Y11/EM)*B3:  B1=B3-B2
160  HN=(B1*H1*H1+B2*H2*H2)/2/(B1*H1+B2*H2):  HM=H2-HN
170  S=B1*H1+B2*H2
180  I=B2*(HN^3+HM^3)/3+B1*H1*(HM-H1/2)^2+B1*H1^3/12
190  LAMDA=PI*(DO+DI)/2/N
200  FREQ=2*PI/LAMDA/LAMDA*SQR(1000*EM*I/RHO/S)
210  '**** PRINT ****
220  PRINT  "B1=";B1,"B2=";B2,"B3=";B3;"mm"
230  PRINT  "H1=";H1,"H2=";H2,"H3=";H3;"mm"
240  PRINT  "HN=";HN,"HM=";HM;"mm"
250  PRINT  "SQUARE MOMENT=";I;"mm4"
260  PRINT  "WAVELENGTH=";LAMDA;"mm"
270  PRINT  "RESONANT FREQUENCY=";FREQ/1000;"kHz"
280  END


RUN
B1= 2.08929    B2= 4.41072    B3= 6.5  mm
H1= 2.5        H2= 3          H3= .5  mm
HN= 1.42925    HM= 1.57075  mm
SQUARE MOMENT= 13.2482  mm4
WAVELENGTH= 18.6751  mm
RESONANT FREQUENCY= 54.5173  kHz
```

FIG. 7.9. Computer program for resonant frequency calculation.

or (5.81) should be analysed. Perhaps it would be more practical instead to accumulate a store of data through numerous design cases which compare approximate design values and measurements.

7.3.2 Dimensions and precision

We now examine how the precision of the stator's dimensions affects the resonant frequency. The area S and its second moment I are given by

$$S = bh \tag{7.11}$$

$$I = bh^3/12. \tag{7.12}$$

To simplify the calculation for I, we have assumed that the metal and ceramic have the same Young's modulus. The frequency is then given by

$$\nu = \frac{\pi h}{\sqrt{3}\lambda^2} \sqrt{\frac{E}{\rho}}. \tag{7.13}$$

The relative change in the resonant frequency with respect to a small variation in height h is given by

$$\frac{\partial \nu}{\partial h} = \frac{\pi}{\sqrt{3}\lambda^2} \sqrt{\frac{E}{\rho}} = \frac{3.14}{1.73 \times 18.7 \times 18.7} \sqrt{\frac{1.12 \times 10^{11}}{8.87 \times 10^3}}$$

$$= 18.5 \times 10^3 \, \text{Hz mm}^{-1} \tag{7.14}$$

while the change in resonant frequency with respect to variation in wavelength λ is given by

$$\frac{\partial \nu}{\partial \lambda} = -\frac{2\pi h}{\sqrt{3}\lambda^3} \sqrt{\frac{E}{\rho}} = -\frac{6.28 \times 3}{1.73 \times 18.7^3} \sqrt{\frac{1.12 \times 10^{11}}{8.78 \times 10^3}}$$

$$= -5.94 \times 10^3 \, \text{Hz mm}^{-1}. \tag{7.15}$$

The wavelength in turn depends on the outer and inner diameters (D_o and D_i) of the stator ring. Since the wavelength is given by eqn (7.8), the variation of wavelength with respect to a change in the outer diameter D_o at a constant inner diameter D_i is given by

$$\frac{\partial \lambda}{\partial D_o} = \frac{\pi}{2m}. \tag{7.16}$$

Therefore

$$\frac{\partial \nu}{\partial D_o} = \frac{\partial \nu}{\partial \lambda} \frac{\partial \lambda}{\partial D_o} = -\frac{\pi^2 h}{\sqrt{3}m\lambda^3} \sqrt{\frac{E}{\rho}}. \tag{7.17}$$

If m, the number of pole pairs, is 9

$$\frac{\partial \nu}{\partial D_o} = -1.04 \times 10^3 \, \text{Hz mm}^{-1}. \tag{7.18}$$

If the variations in D_o and D_i are in the same direction (i.e. they are additive), then

$$\frac{\partial \nu}{\partial D} = -\frac{2\pi^2 h}{\sqrt{3}m\lambda^3} \sqrt{\frac{E}{\rho}} = -2.07 \times 10^3 \, \text{Hz mm}^{-1}. \tag{7.19}$$

From these calculations, we can see that the precision of the ring's dimensions has the following effects on the resonant frequency:

(1) If the stator's thickness h is increased by 0.01 mm, the frequency increases by ~ 185 Hz.

(2) If the stator's outer diameter is increased by 0.01 mm, the frequency decreases by ~ 10 Hz. If both outer and inner diameters are increased by 0.01 mm, the frequency decreases by twice the amount, or 21 Hz.

These dimensional precisions should be taken into consideration when determining the range of source frequency adjustment. Conversely, the dimensional precision must be selected to match the power supply's frequency range. In terms of its effect on resonant frequency, however, temperature change is a more important factor than dimensional precision. The effect of temperature is shown in Fig. 7.10 for a phosphor bronze comb-tooth stator.

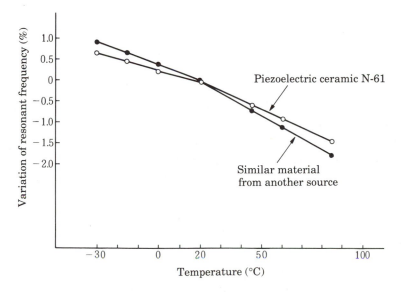

FIG. 7.10. Temperature-dependence of resonant frequency, using a standard motor (type USR60) manufactured by Shinsei Industries Co. Ltd.

However, it is essential to achieve a high level of precision for the flatness and surface roughness of the stator's contact surface with the rotor. From our experience with several test models, a flatness of three to five Newton's rings (or interference fringes) is required (i.e. a height difference of 1–1.5 μm), and a surface of a lapping finish with irregularities $\leqslant 0.4 \mu$m is required.

7.3.3   Force factor

In Section 3.3 we showed that the force factor A_0 of the transverse effect of a ceramic plate, as shown in Fig. 7.11(a), is given by

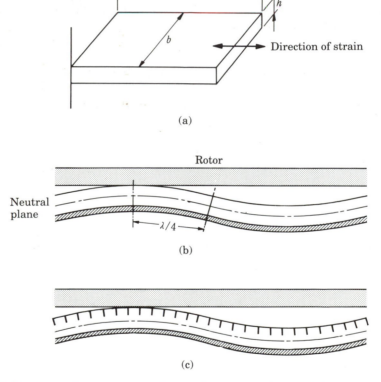

(a)

(b)

(c)

FIG. 7.11.   Force-factor considerations. (a) Force factor $A_0 = bd_{31} Y_{11}$ for transverse effect of a single plate. (b) When the elastic (metal) and ceramic rings vibrate as a single unit, deformation of the ceramic ring and that of the elastic ring's upper surface (in contact with the rotor) are expected to be approximately equal. The plate in (a) corresponds to a quarter-wavelength segment. (c) With the comb-tooth structure, the neutral plane is positioned lower than in (b) and deformation at the stator's contact surface (with the rotor) is greater. Consequently the motor's force factor suffers (is smaller).

$$A_0 = bd_{31} Y_{11} \tag{7.20}$$

where b is the width of the ceramic plate, d_{31} the piezoelectric strain constant, and Y_{11} the Young's modulus.

In this section we derive the force factor A for a complete motor assembly
in terms of A_0, the force factor for a single plate. As noted earlier, the
neutral plane is positioned at approximately the midpoint along the stator's
total thickness h_2 (i.e. thickness of metal and ceramic sheet), and transverse deformations on the opposite surface (i.e. the rotor side) can be considered to be approximately equal to those of the ceramic sheet.

From the analysis developed in Section 5.12 using eqn (5.112), the
single plate shown in Fig. 7.11a is considered to correspond to a quarter-
wavelength segment of the ceramic–metal ring. If we assume that each
phase carries a current I, then each quarter-wavelength segment carries a
current of $I/2(m-1)$. If in eqn (3.11) we assume that the blocking admittance Y_d is cancelled out, then

$$I = Av \tag{7.21}$$

where v, in our case, is the velocity of the expanding or contracting motion
of the quarter-wavelength segment. Since I in this equation represents the
total current, it becomes necessary to convert A into the form

$$A = 2(m-1)A_0. \tag{7.22}$$

In Chapter 6 we proposed using the equivalent circuit shown in Fig. 6.11,
in which the effective (r.m.s.) current in phase A or B is used for analysis.
Thus, dividing by a factor of $\sqrt{2}$, we obtain

$$A = \sqrt{2}(m-1)A_0. \tag{7.23}$$

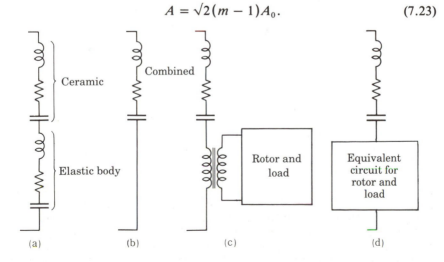

FIG. 7.12. Coupling of the ceramic ring and the elastic ring, and equivalent
circuit representing the comb-tooth effect.

The equivalent circuits in Fig. 7.12 show the coupling of the ceramic and elastic rings. Here (a) shows elements representing the ceramic and elastic rings in series; the elastic ring can be viewed as a load acting on the ceramic ring. The two rings are combined in (b). When the elastic ring has a greater mass as well as a larger spring constant (or a large stiffness), compared with the ceramic ring, the resonant frequency is determined mostly by its properties.

The comb-tooth structure amplifies the stator's motion before it is transmitted to the rotor: this can be viewed as a horn effect. This is represented by the equivalent circuits in Fig. 7.12(c) and (d). For the force factor in this case, eqn (7.23) should be modified with a correction factor ξ to obtain

$$A = \sqrt{2}\xi(m-1)A_0. \tag{7.24}$$

We shall give an example later, in which ξ will be determined experimentally.

7.3.4 No-load speed, starting torque, and operating point for the motor

In an ideal motor with no losses, the starting torque T_{st} and no-load speed N_0 can be expressed in terms of the force factor A as

$$T_{st} = 2RA \cdot V \tag{7.25}$$

$$N_0 = \frac{1}{RA} I_{st} \tag{7.26}$$

where V is the applied voltage, I_{st} is the starting current:

$$I_{st} = V/\sqrt{[(\omega L_m - 1/\omega C_m)^2 + r_0^2]} \tag{7.27}$$

and R is the effective radius:

$$R = (D_o + D_i)/4. \tag{7.28}$$

Under actual conditions, these values would be lower owing to losses. We shall assume that the relation between the speed and torque is given by a straight line connecting the no-load speed and starting torque, as in Fig. 7.13. In motors with relatively low efficiencies, the maximum values for output P_{OUT} and efficiency η are obtained at approximately one-half of the no-load speed. Therefore the load is normally adjusted so that the motor's operating point falls in this vicinity.

7.3.5 Blocking capacitor and cancelling inductance

An inductor can be placed to cancel out the reactive current in a blocking capacitor. To determine this inductance, we need to know the capacitance.

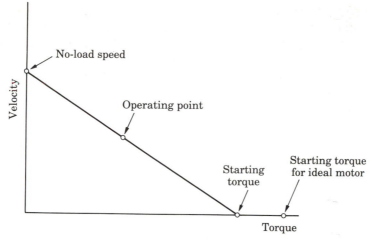

FIG. 7.13.   No-load speed, starting torque and operating point.

In theory, this is given in terms of the piezoelectric ceramic's relative permittivity ε, the effective area S_c for one phase and the thickness h_c, as

$$C_d = \frac{\varepsilon \varepsilon_0 S_c}{h_c}. \tag{7.29}$$

However, in the ultrasonic motor, the value of the dielectric constant is uncertain, since the stator consists of alternately poled segments, and this equation must be used with some caution. To give an example, for Tokin Corporation's NEPEC-61, assuming $\varepsilon = \varepsilon_{33}^T = 1400\varepsilon_0$ for the dielectric constant, we have

$\varepsilon_0 = 8.86 \times 10^{-12}\,\mathrm{F\,m^{-1}}$

$h_c = 0.5\,\mathrm{mm} = 5 \times 10^{-4}\,\mathrm{m}$

$S_c = 6.5\,\mathrm{mm} \times 53.5\,\mathrm{mm} \times (160°/180°) \times 3.14 = 971\,\mathrm{mm^2} = 9.71 \times 10^{-4}\,\mathrm{m^2}$

So we get $C_d = 2.4089 \times 10^{-8}\,\mathrm{F} = 24.09\,\mathrm{nF}$. The measured result, on the other hand, was 9.01 nF.

The cancelling inductance L_c is given by

$$L_c = \frac{1}{(2\pi\nu_0)^2 C_d} \tag{7.30}$$

where ν_0 is the resonant frequency. For $\nu_0 = 45.5\,\mathrm{kHz}$, we obtain $L_c = 1.35\,\mathrm{mH}$. In some circumstances, however, a larger inductance value may be used, as for instance when an excitation frequency slightly higher than the resonant frequency is used. (This will be discussed later.) In this case, the main circuit becomes inductive, so that the compensating inductance can be lower than the value derived from eqn (7.30). The C_d value varies

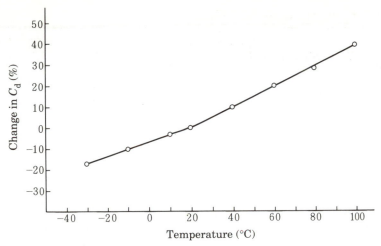

Fɪɢ. 7.14.   Temperature effect on blocking capacitance for NEPEC-61 ceramic.

considerably with temperature (Fig. 7.14), and this must also be taken into consideration.

7.4   Materials

The authors tested various stator materials for an experimental motor without comb tooth structure. The results are presented in this section.

7.4.1   Stator

The stator consists of the elastic ring, the piezoelectric ceramic ring, and an adhesive to bond the two parts together. The selection of materials greatly affects the motor's performance. In these tests, several materials were tested under identical conditions and the results were compared.

(1)   *Elastic ring*. For the elastic ring, we cut thin slices from a $B_S B_M 60/40$ bronze pipe, which we selected for its workability. More recently we have found phosphor bronze to be a more suitable material owing to its superior anti-abrasion properties as well as a nearly uniform Young's modulus insensitive to temperature changes.

(2)   *Piezoelectric ceramic ring*. For the piezoelectric element, we used the NEPEC product series, manufactured by Tokin Corporation. The main component of these ceramics is lead zirconate titanate. The disc, which has an outer diameter of 60 mm and a thickness of 0.5 mm, has silver electrodes baked on to it. From the series, we selected NEPEC-61, which has a high

Table 7.1. Properties of NEPEC series

Property	Symbol	Units	N–10	N–21	N–61
Permittivity	ε_{33}^T	$10^{-8}\,F\,m^{-1}$	4.82	1.59	1.24
	ε_{11}^T		4.42	1.74	1.15
Tangent delta	$\tan\delta$	%	2.0	2.0	0.3
Electromechanical coupling coefficient	k_{31}	non-dimensional	0.34	0.38	0.33
	k_{33}		0.68	0.73	0.67
Young's modulus	Y_{11}	$10^{10}\,N\,m^{-2}$	6.8	6.1	7.6
	Y_{33}		5.5	5.0	6.4
Piezoelectric strain constant	d_{31}	$10^{-10}\,m\,V^{-1}$	−2.87	−1.98	−1.32
	d_{33}		6.35	4.17	2.96
Voltage output constant	g_{31}	$10^{-3}\,V\,m\,N^{-1}$	−5.95	−12.4	−10.6
	g_{33}		13.1	16.2	23.8
Poisson ratio	σ^E	non-dimensional	0.34	0.34	0.31
Mechanical quality factor	Q	non-dimensional	70	75	1800
Curie temperature	T_c	°C	145	330	315
Density	ρ	$10^3\,kg\,m^{-3}$	8.00	7.82	6.90

mechanical quality factor (Q), and NEPEC-10 and NEPEC-21, which have high piezoelectric constants. Their properties are listed in Table 7.1.

(3) *Adhesive*. The adhesive must satisfy as many of the following requirements as possible:

(a) create strong metal-to-metal and metal-to-ceramic bonds;

(b) possess high bonding and peel strengths;

(c) stick by chemical or heat-setting means;

(d) reasonable setting time;

(e) high heat durability.

Some epoxy resins and two types of anaerobic adhesive were tested. First to be tested was an epoxy phenol adhesive, which has good heat resistance properties and a high quality factor Q. However, it required a setting time of 20 min at a temperature of 160°C, and some of the samples developed strains or cracks after application. Therefore we discarded this material. Currently we mostly use high-temperature setting epoxy resins. The three adhesives we tested in the early stages are:

Table 7.2. Loctite test results

Property	Test method	Respresentative values
(a) Hardened strength when applied on mild steel (Loctite No. 648)		
Static shear strength	MIL-R-46082A	25–30 Mpa
Tensile shear strength	ASTM D-1002	12.5–15.5 MPa
Impact value	ASTM D-950	10–15 kPa m
(b) 0.05 mm thick adhesive layer applied on aluminium or steel (Loctite No. 324)		
Tensile shear strength	ASTM D-1002	23 MPa (aluminium) 22.5 MPa (steel)
Tensile strength	ASTM D-2095	35 MPa
Impact value	ASTM D-950	20 kPa m
Peel strength	ASTM D-1876	1.4 MPa (aluminium) 2.7 MPa (steel)

(a) a room-temperature setting epoxy resin (Araldite AW106);

(b) an anaerobic adhesive (Loctite No. 324);

(c) an anaerobic adhesive (Loctite No. 648).

The adhesive strength properties of Loctite 324 and 648 are given in Table 7.2.

7.4.2 Rotor and lining material

In the ultrasonic motor, the oscillation energy generated at the stator's surface must effectively be transmitted to the rotor to create unidirectional motion. Although our understanding of the complex stator–rotor interaction is incomplete, we have found that a thin ring (or lining) between the rotor and stator can be effective. The material for the lining should satisfy the following requirements:

(1) uniform distribution of pressure to the stator's contact area;

(2) elasticity such that the length of each contact area at the wave crests extends from 1/8 to 1/4 wavelength;

(3) A high coefficient of friction;

(4) A high quality factor Q.

Figure 7.15(a) shows a fibrous material with a high Young's modulus used for the lining. The rotor in (b) has protruding flanges so that contact is always made with the stator. In our tests, we used the method shown in (a), with a high polymer for the lining. The dimensions of the rotor are given in Fig. 7.16. Aluminium was used for the rotor.

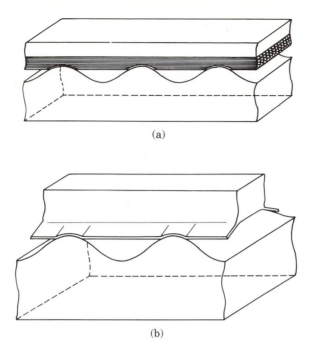

(a)

(b)

Fig. 7.15.   Lining materials: (a) high-polymer lining; (b) flanged lining.

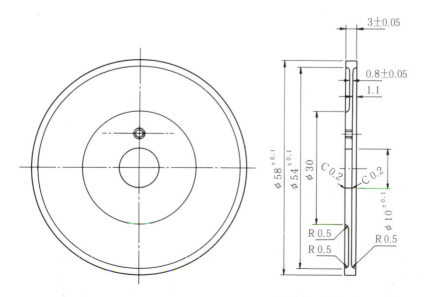

Fig. 7.16.   Rotor dimensions, mm (material: aluminium).

7.5   Testing and evaluation

After narrowing down the choice of materials, a number of motors were constructed, tested, and evaluated. We shall here explain the processes and the results.

7.5.1   Circuit diagram

Figure 7.17 shows the system diagram for our testing apparatus. For torque measurements, we used either the Prony brake or the hysteresis brake method[1]. A photodetector was used for measuring the oscillating amplitude of the stator with and without the rotor. In this sensor, which is highly accurate, light guided through optic fibres illuminates the stator surface. The amount of reflected light is then measured to determine the distance to the surface.

7.5.2   Losses

Numerous factors contribute to losses in the travelling-wave motor, and it is difficult to assess all these factors accurately. It is possible, however, to

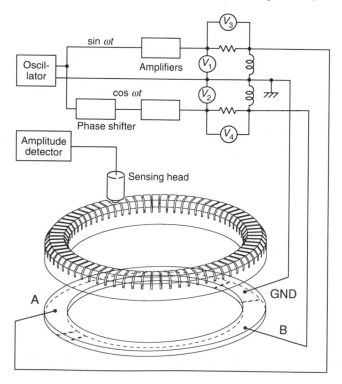

Fɪɢ. 7.17.   Circuit diagram for measurements.

identity a few major ones by making measurements of losses under various conditions.

(1) *Losses due to the stator's unrestrained oscillations*. We took measurements of the electric power required to maintain a transverse-wave amplitude of 0.35 μm on the stator. The results are shown in Table 7.3. The elastic ring had an outer diameter of 50 mm, an inner diameter of 40 mm, and a thickness of 2.5 mm. For the adhesive, Loctite No. 648 was used. The results show that NEPEC-61 had by far the lowest losses. In other words, it has a high quality factor, and the latest motors utilize this product. We then used this ceramic to compare three types of adhesives; all three showed low levels of loss, with no significant differences among them.

(2) *Losses at the felt support*. Travelling waves have no stationary nodes along the stator ring, which makes the problem of supporting it a difficult one. A support structure for the stator should ideally satisfy the following conditions:

(1) it does not affect the stator's oscillation;

(2) no oscillations leak through the support;

(3) there should be minimal oscillations in the support itself;

(4) it should uniformly support all parts of the stator.

Two methods are shown in Fig. 7.18:

(a) The back (ceramic ring) is supported by a vibrational insulator such as felt;

(b) a thin metal plate is connected to the inner periphery of the stator.

In our tests, we measured losses using the felt support. The results are shown in Fig. 7.19. In this test, the stator was sandwiched between two sheets of felt, and the electric power necessary to maintain a transverse wave

Table 7.3. Measurements of power consumption necessary to maintain a constant wave amplitude

Piezoelectric ceramic NEPEC series	Frequency (kHz)	Electrode 1 (V, mA)	Electrode 2 (V, mA)	Power consumption (mW)
No. 10	36.1	5.5, 30	5.5, 27	313.5
No. 21	36.7	8.0, 20	8.0, 24	352
No. 61	38.0	1.25, 11	1.25, 15	32.5

Ceramic thickness = 0.5 mm
Elastic body $B_S B_M$: outer diameter = 50 mm; inner diameter = 30 mm; thickness = 2.5 mm
Adhesive: Loctite No. 648

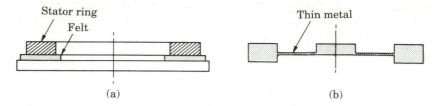

Stator ring

Felt

Thin metal

(a)   (b)

FIG. 7.18.   Stator support methods: (a) using felt; (b) using a metal sheet structure (sectional view).

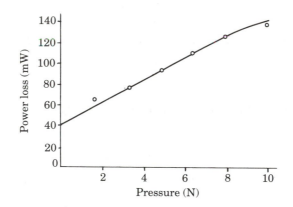

FIG. 7.19.   Applied pressure and losses with felt support.

amplitude of $0.35\,\mu m$ (effective value) was measured for various applied pressures. The results show that the losses are two to three times those obtained when the stator is used without the felt.

7.5.3   Measurement of current–amplitude characteristics and estimation of the force factor

As we showed in Section 3.3, the stator's current is given in terms of the force factor A and blocking admittance Y by

$$I = Av + Y_1 V. \qquad (7.31)$$

Since Y is almost a capacitive reactance, it can be cancelled out by connecting an appropriate inductor in parallel. Thus

$$I = Av. \qquad (7.32)$$

The maximum velocity v is equal to $2\pi\nu u_0$, where u_0 is the oscillation amplitude in the tangential direction at the stator's surface. This velocity is attained when either of the phase currents reaches its peak value (i.e. effective value I_0 multiplied by $\sqrt{2}$). Therefore A can be determined experimentally from the equation

$$A = \frac{\sqrt{2}I_0}{2\pi\nu u_0} = \frac{I_0}{\sqrt{2}\pi\nu u_0}. \tag{7.33}$$

Figure 7.20 shows the relation between transverse-wave amplitude w_0 and the effective current using the elastic ring shown in Fig. 7.8. As we showed in Chapter 5, for a beam with quadrilateral and uniform cross-section, the relation

$$u_0 = \frac{2\pi}{\lambda}\frac{h_2}{2}w_0 \tag{7.34}$$

exists between u_0 and w_0. Instead of $h_2/2$, we shall use h_N to obtain

$$u_0 = \frac{2\pi}{\lambda}h_N w_0. \tag{7.35}$$

If we substitute this equation into eqn (7.33), we get

$$A = \frac{I_0\lambda}{2\sqrt{2}\pi^2\nu h_N w_0}. \tag{7.36}$$

So, substituting values from the linear relation in Fig. 7.20(a), we obtain for A

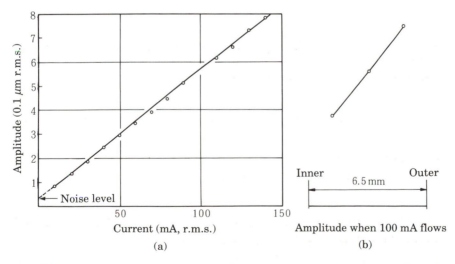

FIG. 7.20. Amplitude–current relation when one phase is excited; the amplitude is greater towards the ring's outer perimeter (C_d was cancelled with a reactor). (a) Measurements taken at the ring's midpoint; (b) position along the ring's width.

$$A = \frac{100 \times 10^{-3} \times 18.7 \times 10^{-3}}{2 \times 1.414 \times 3.14^2 \times 45.23 \times 10^3 \times 1.44 \times 10^{-3} \times 0.54 \times 10^{-6}}$$

$$= 1.907 \, \text{C m}^{-1}. \tag{7.37}$$

Earlier in this chapter, we showed that the force factor A is given by eqn (7.24). On the other hand the force factor A_0 for a single ceramic plate can be determined from eqn (3.26). Thus, using the data in Table 7.1 and ceramic width $b = 6.5 \, \text{mm}$, $A_0 = 6.51 \times 10^{-2} \, \text{C m}^{-1}$. Using $m = 9$, we have

$$A = 1.907 = \xi \times 1.414 \times 8 \times 0.0651 \tag{7.38}$$

From this the correction factor ξ defined in eqn (7.24) is derived as

$$\xi = 2.59. \tag{7.39}$$

7.5.4   Check using the e factor

We can check the relation between current and the ring's oscillation amplitude using the piezoelectric constant and Young's modulus. If we differentiate eqn (3.21) with respect to time, we obtain

$$\frac{d\Delta D}{dt} = e \frac{dS}{dt} + \varepsilon \frac{d\Delta E}{dt}. \tag{7.40}$$

The term on the left is the current per unit area; the terms on the right are, in order, the current caused by the strain in the ceramic and the current through the capacitor (acting as the blocking capacitor). We shall assume that the last term is cancelled out by a reactor. The current per phase can be obtained by multiplying this equation by the area of the ceramic ring:

$$i = (m - 1) B \frac{d\Delta D}{dt} = (m - 1) \, Be \frac{dS}{dt} \tag{7.41}$$

where B is the area of one segment.

To determine the effective value, we can use the relation $d/dt = j2\pi \nu$. Thus

$$I = 2\pi \nu (m - 1) Be \, (\Delta l / l) \tag{7.42}$$

where l is equal to a quarter-wavelength (Fig. 7.11):

$$l = \frac{1}{4} \lambda \tag{7.43}$$

and Δl is the elongation for each ceramic segment. We shall assume that the ratio $\Delta l / \Delta w$ is given by

$$\Delta l / \Delta w = \pi h_2 / \lambda \tag{7.44}$$

from the elliptical trajectory.

Segment area B is given by

$$B = \lambda (D_o - D_i)/4. \tag{7.45}$$

So for the current I we have

$$I = 2\pi^2 e (m - 1) (D_o - D_i) \nu h_2 \Delta w/\lambda. \tag{7.46}$$

Equation (3.25) gives $e = d_{31} Y_{11}$, so we can calculate I from the data in Table 7.1 and the dimensions in Fig. 7.8. In the graph in Fig. 7.20, we get $\Delta w = 0.578\,\mu m$ (effective value) for a current of $100\,mA$. Using the same Δw value, we obtain $I = 87\,mA$ from eqn (7.46).

7.5.5   Stators with comb-tooth structure

Measurement of amplitudes on comb-tooth stator rings is a difficult task; so far, the authors have not been able to obtain definite results. Nor have they established the relation between Δw and Δl. As a general rule, we can expect A to be lower for comb-tooth stators than for flat-plate stator rings (see Fig. 7.11c). In Chapter 6, we used a force factor of $0.24\,C\,m^{-1}$ to calculate the characteristics in Fig. 6.17, assuming the equivalent circuit shown in Fig. 6.14. The theory behind this circuit, however, is still in its developmental stage.

7.6   Test system for use on production lines

The ultrasonic motor has proved to be useful in various industries. Because of its direct-drive quiet operation, it is used for the head support of a VIP seat in limousines. Before this mass-production could take place, technology to speed up the quality-testing of each motor as it was produced had to be available. Manual inspection is too time-consuming, as the characteristics of the ultrasonic motor are sensitive to the finished surfaces of both the stator and the rotor. For fine adjustment and machine automation, we have developed a computer-aided test system[2].

7.6.1   Hardware configuration

Figure 7.21 shows the system configuration. One of the features of this system is that it uses a KENTAC RM 86 board computer, shown in Fig. 7.22, which is designed to be used together with an MS-DOS personal computer serving as the host computer. In order that the capability of the KENTAC could be exploited fully, some hardware was designed, such as a speed/speed-ripple sensing unit, a control signal generator (a bus controller and I/O ports), and a power amplifier. A powder brake was chosen as the load of the test motor. Figure 7.23 shows the test bench.

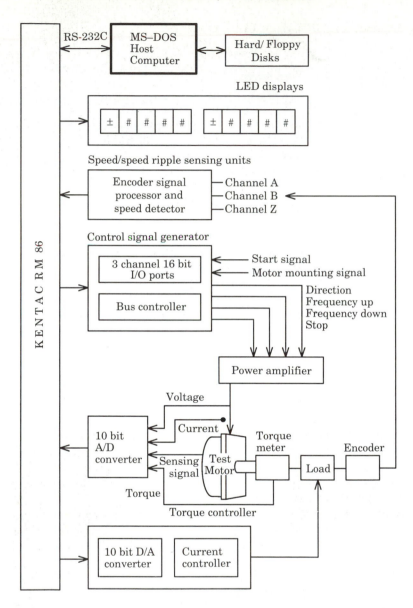

FIG. 7.21. System diagram of the test bench.

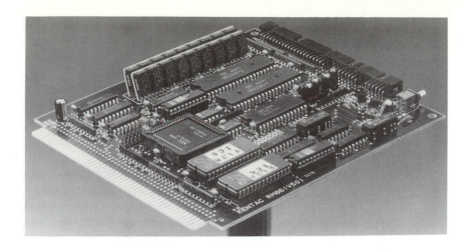

CPU	V50: μPD70216L-8 (8MHz clock no-waiting operation)
Memory	ROM: 27C256x2 (64K bytes)
	D-RAM: HB561003AP – 12x2 (512K bytes)
Parallel I/O	μPD71055Cx2: 2 ports are preset for output, and 2 ports for input, but changeable by user's software
Serial I/O	RS-232C; Baud rate: any at 150 to 19200 bits per second
Interrupt	8 levels
BUS	Based on PC9801 series
Power supply	DC +5V/0.8A
Size	150mm × 170mm × 20mm

Fig. 7.22. KENTAC RM 86 used as the main controller.

Fig. 7.23. Test bench.

7.6.2   Software and operation

The software has been written in C language and is compiled by the host computer. The object language programs thus produced are transmitted via an RS-232C cable to the KENTAC, which controls the measuring system. The measured data are returned to the host computer to be filed in the hard/floppy disks. These files can be used for statistical calculations, e.g. for quality control.

When the system is booted up, the procedure program is automatically transmitted to the RAM area in the KENTAC. In normal operation, a standard drive frequency is first set, at which the initial measurement of torque–speed relationships is done for both directions. When required specifications (for example for speed–torque characteristics and speed ripple) are not satisfied, the drive frequency is readjusted, and testing is pursued until the required specifications are satisfied. Experiment items or specifications can be easily altered and implemented while the production line is carrying out its routine operations. The test that is in progress is always displayed in both the LED panel and the host computer.

References

1. Kenjo, T. (1984). *Stepping motors and their microprocessor controls*, Ch. 6. Oxford University Press.
2. Sashida, T., Kenjo, T. and Takahashi, H. (1992). *Test system designed for use in production lines of ultrasonic motors*, JIEE Technical Report RM-92-21, pp. 117–23.

8. Comparison with electromagnetic motors

In the preceding chapters, we have presented the outline, theory, and practical design aspects of the ultrasonic motor. However, this work is not intended as an exhaustive treatise on the subject and there remain areas in need of further research. In our final chapter, we shall take a different approach and compare the ultrasonic motor with electromagnetic motors, which should highlight the properties and future prospects of the former. It is assumed that the reader is familiar with and possesses a basic knowledge of electromagnetic machines. Reference 1 is recommended for those who have yet to learn about conventional motors.

8.1 Rigid-body theory of electromagnetic motors

In an ultrasonic motor, the stator must deform to maintain frictional contact with the rotor and transmit torque. The components of an ideal electromagnetic motor, however, are rigid bodies and no deformation is assumed to occur. The distinguishing feature is that an air gap exists between the stator and rotor, and rotation occurs without any contact between the two. We shall first see what this premise (of rigidity) means from the standpoint of physics.

8.1.1 Basic theory of electromagnetism

The force acting on a body placed in an electromagnetic field in free space is known to be given by the volume integral (Fig. 8.1)

$$F = \int_V (f_M + f_E)\, dV. \tag{8.1}$$

According to the theory developed by Livens[2], the magnetic volume force f_M and the electrical volume force f_E are expressed as

$$f_M = I \times B + \nabla(M \cdot H) \tag{8.2}$$

$$f_E = \rho E + (P \cdot \nabla)E \tag{8.3}$$

where the ∇ operation does not affect M. Here, the body is assumed to be rigid and hence its deformation is not considered in this equation. There is little advantage to be gained from incorporating deformation into the

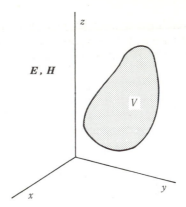

Fig. 8.1. A body in an electromagnetic field.

theory for this case, because there is no contact between rotor and stator.
The symbols used in these equations are defined below.

ρ = electric charge density

E = electric field intensity

H = magnetic field intensity

B = magnetic flux density

D = electric displacement or flux density, which does not appear explicitly
in the above equations

M = magnetization

P = electric polarization

All, except ρ are vector quantities.

Moreover, the following relationships exist among magnetic and electric
quantities;

$$B = M + \mu_0 H \qquad (8.4)$$

$$D = P + \varepsilon_0 E \qquad (8.5)$$

where ε_0 is the permittivity of free space and μ_0 the permeability of free
space.

Depending on the medium, either or both of M and P can be zero (or
sufficiently small to be considered zero). For example, both are considered
zero in air. In vacuum they are absolutely zero.

Equation (8.4) implies that the magnetic flux density in an object is the
sum of the portion $\mu_0 H$ due to the free-space property and the magnetization M of the object. A similar interpretation holds for the electric flux
density given by eqn (8.5).

In general, the directions of the vectors B and H (or D and E) do not coincide. It is then customary to express these quantities as

$$\left. \begin{aligned} B &= \mu \cdot H \\ D &= \varepsilon \cdot E \end{aligned} \right\} \tag{8.6}$$

in which μ and ε are tensor quantities, each with nine elements. This can also be written in the following form:

$$\begin{bmatrix} B_x \\ B_y \\ B_z \end{bmatrix} + \begin{bmatrix} \mu_{xx} & \mu_{xy} & \mu_{xz} \\ \mu_{yz} & \mu_{yy} & \mu_{yz} \\ \mu_{zx} & \mu_{zy} & \mu_{zz} \end{bmatrix} \begin{bmatrix} H_x \\ H_y \\ H_z \end{bmatrix}. \tag{8.7}$$

(Since the equation for electric quantities are similar to the ones for magnetic quantities, we shall omit the former.)

Equation (8.7) can be written in the form

$$\begin{bmatrix} B_x \\ B_y \\ B_z \end{bmatrix} = \begin{bmatrix} (\mu_{xx} - \mu_0) & \mu_{xy} & \mu_{xz} \\ \mu_{yz} & (\mu_{yy} - \mu_0) & \mu_{yz} \\ \mu_{zx} & \mu_{zy} & (\mu_{zz} - \mu_0) \end{bmatrix} \begin{bmatrix} H_x \\ H_y \\ H_z \end{bmatrix} + \mu_0 \begin{bmatrix} H_x \\ H_y \\ H_z \end{bmatrix}. \tag{8.8}$$

Therefore

$$M = \begin{bmatrix} (\mu_{xx} - \mu_0) & \mu_{xy} & \mu_{xz} \\ \mu_{yx} & (\mu_{yy} - \mu_0) & \mu_{yz} \\ \mu_{zx} & \mu_{zy} & (\mu_{zz} - \mu_0) \end{bmatrix} \begin{bmatrix} H_x \\ H_y \\ H_z \end{bmatrix}. \tag{8.9}$$

In general, and in electric machinery, the quantities B, H, D and E are functions of time (t) and position. Furthermore, when hysteresis exists, the B–H relation at some point in time also depends on its previous behaviour, making it a very complex function of time.

Note that eqn (8.1) contains neither μ or ε; i.e. no restrictions are imposed on either constant. This means that regardless of the form of these tensors (to indicate saturation, magnetic anisotropy, hysteresis, etc.), computations can be performed using eqn (8.1). As it turns out, this has considerable significance.

When μ (or ε) at each point in the medium is unaffected by B (or D) and stays constant, the medium is referred as linear.

8.1.2 Electromagnetic and electrostatic motors

Equation 8.1 contains the following four terms:

$I \times B$:   force generated by the interaction between current and magnetic field. The moving-coil motor and eddy-current motor use this principle to create torque.

$\nabla(M \cdot H)$: torque generated by interaction between magnetic polarization (or magnetization, precisely) and the magnetic field. The hysteresis motor, reluctance motor, permanent-magnet a.c. motor and d.c. motor utilize this effect. In fact, the squirrel-cage induction motor, salient-pole synchronous induction motor and slotted d.c motor also belong in this category. This means that this is a most useful electromagnetic force.

ρE: force exerted on a charge by the electric field.

$(P \cdot \nabla)E$: force generated by the interaction between electric polarization and the electric field.

Forces represented by the last two terms are used in an electrostatic motor. However, in practice, the scale of torque generated by such a motor is minuscule in comparison with torque created magnetically. Thus a motor designed to transmit electrostatic forces via an air gap by the use of a high-permittivity material cannot compete in power with a regular electromagnetic motor.

We next examine properties of the electromagnetic motor.

8.1.3 Force due to magnetization

The terms $I \times B$ and $\nabla(M \cdot H)$ represent forces which we shall call the electrodynamic and the magnetization force respectively. The magnetization force, in terms of its components, is given by

$$f_x = M_x \frac{\partial H_x}{\partial x} + M_y \frac{\partial H_y}{\partial x} + M_z \frac{\partial H_z}{\partial x} \tag{8.10a}$$

$$f_y = M_x \frac{\partial H_x}{\partial y} + M_y \frac{\partial H_y}{\partial y} + M_z \frac{\partial H_z}{\partial y} \tag{8.10b}$$

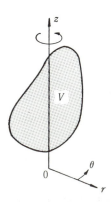

FIG. 8.2. Rotation about the z-axis.

$$f_z = M_x \frac{\partial H_x}{\partial z} + M_y \frac{\partial H_y}{\partial z} + M_z \frac{\partial H_z}{\partial z}. \tag{8.10c}$$

Now, if the movement of a body (a rotor for instance) is limited to rotation around the z-axis (see Fig. 8.2), the torque can be obtained by the volume integral

$$T = \int_V \left\{ r(I \times B)_\theta + M \cdot \frac{\partial H}{\partial \theta} \right\} dV \tag{8.11}$$

where θ is the circumferential coordinate, the subscript indicating that the term is the circumferential component.

8.1.4 Non-salient-pole rotor

In most motors, the rotor is cylindrical. Rotors can be non-cylindrical but are always axially symmetric, or in other words its cross-section normal to the rotating axis is always a circle, as shown in Fig. 8.3. In the terminology of electric machinery, these are called non-salient-pole types. Figure 8.3(d) shows a general axially symmetric body, which we shall consider first. In a non-salient-pole rotor, the torque created by magnetization can be written, substituting B for M, as

$$T_M = \int_V \left(B \cdot \frac{\partial H}{\partial \theta} \right) dV \tag{8.12}$$

which is a more convenient form when analysing forces on the rotor. We shall first present a mathematical proof of this equation.

Proof. A non-salient-pole rotor can be considered to consist of rings as shown in Fig. 8.4, where each ring is an axially symmetric body. Then, $dV = r\,d\theta\,dr\,dz$ in eqn (8.12) is the ring's volume. The volume integral can then be expressed as

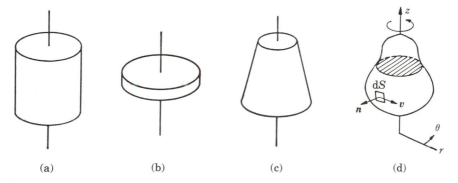

(a)   (b)   (c)   (d)

Fig. 8.3.   Non-salient-pole rotors: (a) cylindrical; (b) disc, (c) tapered; (d) general axially symmetric body of rotation.

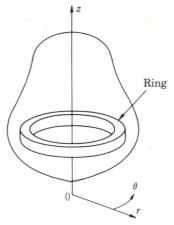

FIG. 8.4.   Non-salient-pole rotor as a collection of rings.

$$T_{\mathrm{M}} = \int_r \int_z \int_{\theta=0}^{2\pi} \left(rM \cdot \frac{\partial H}{\partial \theta} \right) \mathrm{d}\theta \, \mathrm{d}r \, \mathrm{d}z. \tag{8.13}$$

(It is vital to our proof that the integration be carried out from $\theta = 0$ to $\theta = 2\pi$. If the rotor is not axially symmetric, as in Fig. 8.5, then this integration cannot take place.) Consider only the integral with respect to θ in eqn (8.13). Equating it with I, we have

$$I = \int_0^{2\pi} rM \cdot \frac{\partial H}{\partial \theta} \, \mathrm{d}\theta. \tag{8.14}$$

If we substitute

$$M = B - \mu_0 H$$

which is derived from eqn (8.4), into eqn (8.14), we obtain

$$I = \int_0^{2\pi} rB \cdot \frac{\partial H}{\partial \theta} \, \mathrm{d}\theta - \int_0^{2\pi} r\mu_0 H \cdot \frac{\partial H}{\partial \theta} \, \mathrm{d}\theta. \tag{8.15}$$

However, the second term is cancelled since

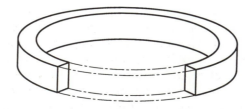

FIG. 8.5.   Incomplete ring.

$$-\int_0^{2\pi} r\mu_0 \boldsymbol{H} \cdot \frac{\partial \boldsymbol{H}}{\partial \theta} \, d\theta = \frac{-r\mu_0}{2} \int_0^{2\pi} \frac{\partial H^2}{\partial \theta} \, d\theta = \frac{-r\mu_0}{2} \int_0^{2\pi} dH^2$$

$$= \frac{-r\mu_0}{2} \{ [H^2]_{\theta=0} - [H^2]_{\theta=2\pi} \} = 0. \quad (8.16)$$

Hence only the first term remains in eqn (8.15). Equation (8.12) follows directly.

A rotor with recessed or concave parts such as slots can be considered axially symmetric if the gaps are filled with, say, resin. Equation (8.16) would then still apply.

8.2   Torque in terms of surface forces

The ultrasonic motor employs surface forces produced at the rotor–stator contact surface. The electromagnetic motor can also be considered in terms of surface forces.

8.2.1   *Volume force and surface force*

Equation (8.17) below represents the torque acting on a rotor. There is a another, simpler form, which is the surface integral of eqn (8.18).

$$T = \int_V \left\{ r(\boldsymbol{I} \times \boldsymbol{B})_\theta + \boldsymbol{B} \cdot \frac{\partial \boldsymbol{H}}{\partial \theta} \right\} dV \qquad \text{(volume integral)} \qquad (8.17)$$

$$= \int_S r H_\theta \boldsymbol{B} \cdot \boldsymbol{n} \, dS \qquad \text{(surface integral)} \qquad (8.18)$$

where S is the surface area of the rotor, $\boldsymbol{n}$ is unit vector at any point on surface S which is normal to the surface ($\boldsymbol{B} \cdot \boldsymbol{n}$ is the scalar product of vectors $\boldsymbol{B}$ and $\boldsymbol{n}$), and r (in eqn 8.18) is the distance between the z-axis and surface S.

The surface integral above can be interpreted as follows. A tangential force (i.e. force in the θ-direction) of $H_\theta \boldsymbol{B} \cdot \boldsymbol{n}$ acts at each point on the rotor's surface. This force is multiplied by r to obtain torque. We then integrate these torques over the rotor's surface to obtain the total torque.

The volume integral, on the other hand, is based on the recognition that torque acts on individual volume elements of the rotor. However, this must not be taken to mean that the two types of forces, surface and volume, simultaneously act on the rotor; rather, only one type is considered effective, depending on which interpretation is taken. This is closely related to the assumption of a rigid body for the rotor stated earlier.

Our theory is based on the assumption that the rotor is a rigid body. In rigid-body mechanics, one can measure forces (or torque) acting on the body as a whole, but not those acting on a part of the body. This is because forces at some part can be determined by deformation or strain measured at that part, and this is not available for a rigid body. Hence, we may assume that the total force or torque acting on the rotor is either the integration of the forces acting on each part of the rotor, or instead the integration of the forces acting only on the rotor surface. Physically speaking, the rotor experiences a volume force, however, so the latter is merely a theoretical interpretation. We shall now derive eqn (8.18).

Proof. We first note that

$$(I \times B)_\theta = I_z B_r - I_r B_z. \tag{8.19}$$

The frequency of the electromagnetic field is so low in a normal motor that the displacement current $\partial D/\partial t$ can be ignored. Hence the current I is given by

$$I = \text{rot } H. \tag{8.20}$$

The components of rot H in cylindrical coordinates are given by

$$(\text{rot } H)_z = \frac{1}{r} \left\{ \frac{\partial (rH_\theta)}{\partial r} - \frac{\partial H_r}{\partial \theta} \right\} \tag{8.21}$$

$$(\text{rot } H)_r = \frac{1}{r} \frac{\partial H_z}{\partial \theta} - \frac{\partial H_\theta}{\partial z} \tag{8.22}$$

$$(\text{rot } H)_\theta = \frac{\partial H_r}{\partial z} - \frac{\partial H_z}{\partial r} . \tag{8.23}$$

Therefore

$$r(I \times B)_\theta = \left\{ \frac{\partial (rH_\theta)}{\partial r} - \frac{\partial H_r}{\partial \theta} \right\} B_r - \left(\frac{\partial H_z}{\partial \theta} - r \frac{\partial H_v}{\partial z} \right) B_z. \tag{8.24}$$

On the other hand,

$$B \cdot \frac{\partial H}{\partial \theta} = \frac{\partial (B \cdot H)}{\partial \theta} - H_\theta \frac{\partial B_\theta}{\partial \theta} - H_r \frac{\partial B_r}{\partial \theta} - H_z \frac{\partial B_z}{\partial \theta} . \tag{8.25}$$

From div $B = 0$, expressed in cylindrical coordinates, we obtain

$$\frac{\partial B_\theta}{\partial \theta} = - \frac{\partial (rB_r)}{\partial r} - r \frac{\partial B_z}{\partial z} . \tag{8.26}$$

Substituting this into the second term of the right-hand side of eqn (8.25) yields

$$B \cdot \frac{\partial H}{\partial \theta} = \frac{\partial (B \cdot H)}{\partial \theta} + H_\theta \left\{ \frac{\partial (rB_r)}{\partial r} + r \frac{\partial B_r}{\partial z} \right\} - H_r \frac{\partial B_r}{\partial \theta} - H_z \frac{\partial B_z}{\partial \theta} . \tag{8.27}$$

The integrand of eqn (8.17) is the sum of eqns (8.24) and (8.27). Organizing the terms for the derivatives with respect to variables r, θ and z, we obtain the following equation:

$$T = \int_V \left\{ \frac{1}{r} \frac{\partial (r H_\theta B_\theta)}{\partial \theta} + \frac{1}{r} \frac{\partial (r^2 H_\theta B_r)}{\partial r} + \frac{\partial (r H_\theta B_z)}{\partial z} \right\} dV. \quad (8.28)$$

Since the integrand takes the form of the divergence of vector $r H_\theta B$, this is equal to the surface integral in eqn (8.18), from Gauss's theorem. QED.

8.2.2 Torque expressed in terms of magnetic flux inclination at the rotor surface

To see what eqn (8.18) means physically, we shall consider a simple cylindrical rotor (see Fig. 8.6). At the side, the unit normal vector n has only the radial component n_r, while at the top and bottom surfaces it has only the z-component n_z. Therefore

$$\boldsymbol{B} \cdot \boldsymbol{n} = \begin{cases} B_r & \text{for the side} \\ B_z & \text{for the ends.} \end{cases} \quad (8.29)$$

So eqn (8.18) becomes

$$T = \int_{\text{side}} R H_\theta B_r \, dS + \int_{\text{ends}} r H_\theta B_z \, dS. \quad (8.30)$$

As shown in the figure, this means that a torque of $R H_\theta B_r$ acts along the side, while a torque of $r H_\theta B_z$ acts at the top and bottom surfaces. In regular motors with cylindrical rotors, the drive is mainly provided by the torque acting along the side, while in disc-shaped motors (e.g. printed-circuit motors), this is supplied by torque acting on the top and bottom surfaces.

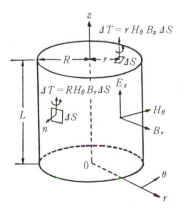

FIG. 8.6.   Torque acting on a cylindrical rotor.

8.3   Analysis using the Poynting vector

Having expressed torque in terms of surface forces above, we shall now use a similar method to derive the motor's output and losses.

8.3.1   The Poynting vector $E \times H$

When electromagnetic waves are emitted from a transmitting antenna to be received by another antenna, the waves carry not only information but also energy. The flow of this energy is expressed by the Poynting vector, which is expressed by $P = E \times H$ or the vector product of the electric field strength E and magnetic field strength H. The magnitude of the Poynting vector is given by the amount of energy that passes through a unit area normal to the vector per unit time, as shown in Fig. 8.7.

Energy is transferred by the Poynting vector in the airgap of an electro-magnetic motor as well. Although no current flows between the stator and rotor by physical contact in a brushless d.c. motor or a.c. motor, electromagnetic flow occurs between the two via the airgap. Currents in the stator windings create a rotating magnetic field, which is a form of electromagnetic wave carrying energy to the rotor via the airgap. Part of the energy is converted into heat within the rotor, and the remainder is converted into mechanical energy (i.e. mechanical work or output).

Let the non-salient-pole body in Fig. 8.3(d) be the rotor. First we shall assume that the rotor is not moving. The energy P_{IN} flowing from the stator (which is not shown in the figure) into the rotor via the airgap per unit time is given by the Poynting vector as follows:

$$P_{\text{IN}} = \int_{S} - (E \times H) \cdot n \, dS. \tag{8.31}$$

The minus sign is necessary since the unit normal vector n points outward from the surface. Values for E and H can be chosen at either side of surface S, because $E \times H$ is continuous at the boundary surface.

Since the rotor is stationary and does not produce any mechanical output, P_{IN} is thought to be converted into heat loss and an increase in electro-

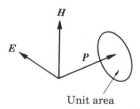

Unit area

FIG. 8.7.   The Poynting vector.

magnetic energy within the rotor. Then Maxwell's equation and Gauss's theorem can be used to show (which details we omit) that the above equation becomes

$$P_{IN} = \int_V \left(E \cdot I + H \cdot \frac{\partial B}{\partial t} + E \cdot \frac{\partial D}{\partial t} \right) dV. \tag{8.32}$$

The three terms of the integrand are explained below.

$E \cdot I$: the Joule loss per unit volume and time

$H \cdot \dfrac{\partial B}{\partial t}$: the increase in magnetic field energy per unit volume and time (when magnetic hysteresis exists, the hysteresis loss is included)

$E \cdot \dfrac{\partial D}{\partial t}$: the increase in electric field energy per unit volume and time, including the heat loss caused by static hysteresis if it exists; this is the same type of energy as the electrostatic energy stored in a blocking capacitance, discussed in Section 3.7.

This is known as Poynting's theorem and can be found in any college textbook on electromagnetism. It should be noted that this theorem itself is not restricted to axially symmetric bodies, although the assumption of rigidity might be required.

8.3.2 Poynting's theorem applied to a rotating body

We shall now examine the case of a moving rotor. In the literature there is very little which applies Poynting's theorem to moving bodies. Although Poynting himself published one such paper[3], his treatment is somewhat questionable in view of current theory. We shall therefore introduce a simple treatment developed by one of the authors[4].

(1) *Rotor input.* Since the rotor is assumed to be axially symmetric, there is no distinction as to whether the rotor is turning or not when observed from outside, and so the energy inflow to the rotor can be given by eqn (8.31).

(2) *Rotor loss.* We now determine the Joule loss and the magnetic/electrostatic hysteresis within the rotor, using a coordinate system which is fixed to (and rotates with) the rotor. Since the rotor is non-moving in this coordinate system, Poynting's theorem may apply. Therefore

$$P_{LOSS} = \int_S - (E' \times H') \cdot n \, dS \tag{8.33}$$

where the dash ($'$) indicates quantities observed from the rotor's coordinate system.

Although this equation includes the energy increase in the electromagnetic field as well as loss, the latter is more important in motor theory. This is because the loss will accumulate with increasing time, whereas the change in electromagnetic energy alternates between positive and negative values so that its average approximates to zero.

(3) *Mechanical output*. Mechanical work (or output) should be the difference between P_{IN} and P_{LOSS}. We shall find out below.

$$P_{OUT} = P_{IN} - P_{LOSS} = \int_{S} (E \times H - E' \times H') \cdot n \, dS. \qquad (8.34)$$

The following relations exist between electromagnetic quantities in the stator and rotor coordinate systems.

$$E' = E + v \times B \qquad (8.35)$$

$$H' = H - v \times D \qquad (8.36)$$

where v is the velocity at the rotor surface. Hence

$$E' \times H' = E \times H + (v \times D) \times E + (v \times B) \times H + \{v \cdot (D \times B)\}v \qquad (8.37)$$

and

$$P_{OUT} = \int_{S} \{(v \times B) \times H + (v \times D) \times E\} \cdot n \, dS$$

$$+ \int_{S} \{v \cdot (D \times B)\} (v \cdot n) \, dS. \qquad (8.38)$$

From our assumption of an axially symmetric body, v and n are perpendicular to each other at the surface so, $v \cdot n = 0$. Thus the second term on the right-hand side above is zero. Furthermore, from formulae in vector analysis, we obtain

$$\{(v \times B) \times H\} \cdot n = (H \cdot v)(B \cdot n) - (H \cdot B)(v \cdot n) \qquad (8.39)$$

$$\{(v \times D) \times E\} \cdot n = (E \cdot v)(D \cdot n) - (E \cdot D)(v \cdot n) \qquad (8.40)$$

The terms $v \cdot n$ which appear in these equations also disappear when integration is carried out. Therefore

$$P_{OUT} = \int_{S} \{(H \cdot v)(B \cdot n) + (E \cdot v)(D \cdot n)\} \, dS. \qquad (8.41)$$

For rotation about the z-axis, v consists of only the tangential component, which is equal to rN (N is the rotational speed in rad s^{-1}). Therefore

$$H \cdot v = rNH_{\theta} \qquad (8.42)$$

$$E \cdot v = rNE_{\theta}. \qquad (8.43)$$

Substituting these into eqn (8.42) yields the following result.

$$P_{OUT} = N \int_S r(H_\theta B + E_\theta D) \cdot n \, dS. \tag{8.44}$$

8.3.3 Expression for torque including electrostatic forces

Torque T times rotational speed N equals mechanical output. Or,

$$P_{OUT} = NT. \tag{8.45}$$

Therefore, using eqn (8.44), torque is given by

$$T = \int_S r(H_\theta B + E_\theta D) \cdot n \, dS. \tag{8.46}$$

The first term of the integrand is the magnetic torque and agrees with eqn (8.18), while the second term is the electrostatic torque. We see thus that torque results from electric and magnetic lines of force. In practice, it is difficult to produce an effective torque from electric lines of force.

This can be seen from the viewpoint of eqn (8.36) as well. This equation indicates that the magnetic field strength H depends on the coordinate system being used. (Equation (8.35) indicates that the electric field strength depends on the coordinate system, which is the well-known Fleming's right-hand rule.) Specifically, eqn (8.36) states that a body travelling at velocity v through an electric flux density D experiences a magnetic field strength $v \times D$. This is a negligible amount, however, and if we use $H' = H$ instead of eqn (8.36), the term for electrostatic torque in eqn (8.46) will disappear.

8.3.4 Energy vector which transmits mechanical energy

Equation (8.34), which we proved above, indicates that mechanical energy is propagated by the electromagnetic vector $E \times H - E' \times H'$. Note however that this is valid only if this vector is integrated over the surface of a cylindrical or other non-salient-pole rotor.

8.4 Analysis of magnetization torque using Stieltjes integrals

Although there are many texts that discuss the torque created by electro-dynamic force, there are few that give adequate explanations of the magnetization torque. Yet the hysteresis, reluctance, permanent-magnet and stepping motors are all applications of the magnetization torque, and are widely used in the electronics field as small control motors. In this section we examine the theory of torque generation in these motors and introduce a method for computing the torque. The method we present was originally developed as the theory of the hysteresis motor by B. R. Teare[5,6] and was extended by one of the authors (T. K.) so that it could be applied to other motors as well. We shall then use this method to make a comparative analysis with the ultrasonic motor.

In the motor, B and H are functions of time t and the circumferential coordinate θ. (They are also functions of r and z, but in the following discussion t and θ will be the relevant variables). Viewed from another angle, we can say that B is a function of H. This relationship becomes important when magnetic saturation or hysteresis occurs in the core. Thus we need to establish how these functions relate to one another. First we make the distinction between the independent variables t and θ and the dependent variables H and B. When B is a function of H, we then consider the dependent variable B as a function of the other dependent variable H through the independent variable t or θ as a parameter.

To examine the magnetization torque in eqn (8.12), we shall express it in the form of a Stieltjes integral.

8.4.1 The Stieltjes integral

First we shall explain what the Stieltjes integral is, although we shall not be concerned with mathematical rigour here.

(1) *For a single variable.* Consider the functions $f(x)$ and $g(x)$ defined on the interval $[a, b]$ (i.e. $f(x)$ and $g(x)$ are the dependent variables, while x is the independent variable). Then the integrals defined by

$$\int_{x=a}^{b} f(x)\, \mathrm{d}g(x), \quad \int_{x=a}^{b} g(x)\, \mathrm{d}f(x)$$

are known as the Stieltjes integrals. In other words, the Stieltjes integral is obtained when one dependent variable is integrated with respect to another dependent variable. A regular Riemann integral, on the other hand, is the integral with respect to an independent variable. These two integrals are related thus:

$$(\mathrm{S})\int_{x=a}^{b} f(x)\, \mathrm{d}g(x) = (\mathrm{R})\int_{a}^{b} f(x)\, \frac{\mathrm{d}g(x)}{\mathrm{d}x}\, \mathrm{d}x. \qquad (8.47)$$

(2) *The Stieltjes integral for a periodic function.* Let $f(x)$ and $g(x)$ be periodic functions with a common period x_0. For any arbitrary x_1, we shall take the integrals for one period (or cycle),

$$\int_{x_1}^{x_1 + x_0} f(x)\, \mathrm{d}g(x), \quad \int_{x_1}^{x_1 + x_0} g(x)\, \mathrm{d}f(x) \qquad (8.48)$$

and express them as

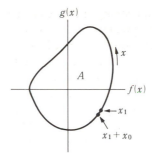

FIG. 8.8.   The $\{f(x), g(x)\}$ plane.

$$\oint f(x)\,\mathrm{d}g(x), \quad \oint g(x)\,\mathrm{d}f(x) \qquad (8.49)$$

respectively. We call these the periodic Stieltjes integrals. If we plot $g(x)$ and $f(x)$ on the ordinate and abscissa respectively, as shown in Fig. 8.8, a locus is obtained. Letting A denote the area enclosed by this loop, the following relation is obtained:

$$\oint f(x)\,\mathrm{d}g(x) = -\oint g(x)\,\mathrm{d}f(x) = \begin{cases} A \text{ (when integration is taken in} \\ \quad \text{the counterclockwise direction)} \\ -A \text{ (when integration is taken in} \\ \quad \text{the clockwise direction).} \end{cases}$$

$$(8.50)$$

For example, if B and H are periodic functions of t and display a hysteresis loop as shown in Fig. 8.9(a), $\oint H\,\mathrm{d}B$ is equal to the area within the loop and corresponds to the hysteresis loss for one cycle. If, as in (b), a minor loop appears within the major one, the area of the minor loop (hatched area) is added to the major loop area. When the minor loop becomes as

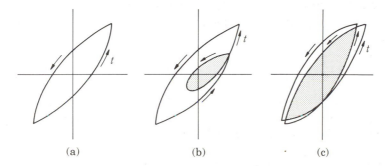

(a)   (b)   (c)

FIG. 8.9.   Examples of hysteresis curves: (a) major loop only; (b) one minor loop; (c) minor loop with increased area.

large, as shown in (c), the computed area is approximately twice the size of the major loop area. The two loops will exactly coincide when the integration comprises two cycles in effect, as for a four-pole motor.

(3)   *Stieltjes integral for two variables.* We now define $f(x, y)$ and $g(x, y)$ with x and y as independent variables. If f and g are periodic functions with periods x_0 and y_0, the meaning of the expression $\oint f \, dg$ as the periodic Stieltjes integral is unclear. In this case, there are two possibilities, as shown in relation to the Riemann integrals below:

$$\left.\begin{aligned} (S) \oint_x f \, dg &= (R) \int_{x_1}^{x_1 + x_0} f \frac{\partial g}{\partial x} \, dx \\ (S) \oint_y f \, dg &= (R) \int_{y_1}^{y_1 + y_0} f \frac{\partial g}{\partial y} \, dy \end{aligned}\right\}. \tag{8.51}$$

The first integral is taken with respect to x, and is referred to as the 'periodic integral with parameter x'. The second one is the 'periodic integral with parameter y'. Although these two are equal in some special cases, they are not equal in general.

8.4.2 Applications of the periodic Stieltjes integral

The periodic Stieltjes integral is useful for expressing physical quantities as follows.

(1)   *Torque.* The magnetization torque for an axially symmetric rotor is given by

$$T = \int_z \int_r r \left(\int_0^{2\pi} \boldsymbol{B} \cdot \frac{\partial \boldsymbol{H}}{\partial \theta} \, d\theta \right) dr \, dz. \tag{8.52}$$

If $2p$ is the number of poles, the angle covered by one cycle is $2\pi/p$. Therefore

$$T = p \int_z \int_r r \left(\int_0^{2\pi/p} \boldsymbol{B} \cdot \frac{\partial \boldsymbol{H}}{\partial \theta} \, d\theta \right) dr \, dz. \tag{8.53}$$

The integral in the parentheses can be rewritten using the relationship in eqn (8.51) to obtain

$$T = p \int_z \int_r r \left(\oint_\theta \boldsymbol{B} \cdot d\boldsymbol{H} \right) dr \, dz. \tag{8.54}$$

(2)   *Rotor input.* Assuming the current within the rotor to be zero and ignoring the term for electrostatic energy in eqn (8.32), the energy inflow to the rotor is given by the volume integral

$$P_{\text{IN}} = \int_V \boldsymbol{H} \cdot \frac{\partial \boldsymbol{B}}{\partial t} \, dV. \tag{8.55}$$

If we assume that both H and B fluctuate with a period t_0, the average value P_{IN} over time is given by

$$\langle P_{\text{IN}} \rangle = \int_V \frac{1}{t_0} \int_{t_1}^{t_1 + t_0} H \cdot \frac{\partial B}{\partial t} \, dt \, dV. \qquad (8.56)$$

Rewriting this using eqn (8.51), we obtain

$$P_{\text{IN}} = \int_V \frac{1}{t_0} \oint_t H \cdot dB \, dV \qquad (8.57)$$

(the angle brackets are omitted in this and subsequent equations).

(3) *Rotor loss*. To determine rotor loss, we perform the above calculation in the rotor or dashed ($'$) coordinate system, as we did in the section on Poynting vectors. Here, for simplicity we shall assume that B and H remain the same in both coordinate systems. Therefore

$$P_{\text{LOSS}} = \int_V H \cdot \left(\frac{\partial B}{\partial t} \right)' dV \qquad (8.58)$$

where ( )$'$ indicates that the derivative is taken with respect to the rotor coordinate system. Thus, similarly to above,

$$P_{\text{LOSS}} = \int_V \frac{1}{t_0'} \int_{t'}^{t' + t_0'} H \cdot \left(\frac{\partial B}{\partial t} \right)' dt \, dV$$

$$= \int_V \frac{1}{t_0'} \oint_{t'} H \cdot d B \, dV. \qquad (8.59)$$

Note that t_0', the period viewed from the rotor coordinate system, is not the same as t_0 in the stator system. The subscript t' in eqn (8.59) indicates that the integration for parameter t is taken with respect to the rotor coordinate system.

We have shown that torque, rotor input and loss in the rotor can all be expressed by the integral $\oint H \cdot dB$ for a motor which utilizes only magnetization for its drive. We shall now examine the hysteresis motor using this method, and then compare it with the ultrasonic motor.

8.4.3 Analysis of a hysteresis motor

Before proceeding, we should briefly explain what a hysteresis motor is. A hysteresis motor is an a.c. motor which uses a permanent-magnet material with a low coercive force for the main rotor. It requires no auxiliary start-up mechanism, accelerating smoothly to enter synchronized operation with the rotating magnetic field. It is thus a type of synchronous motor, and in the past has found wide use as the multispeed capstan motor for tape recorders and has been produced in large quantities. Figure 8.10 shows

FIG. 8.10. Stator and cast-magnet rotor rings for the hysteresis motor.

a stator of such a hysteresis motor and cast magnetic rings used in its rotor. The ring is often referred to as a hysteresis ring.

However, despite its rugged construction and simple operation, the hysteresis motor was found lacking in the precision of rotation necessary for the rotating drums in video tape recorders. Because of this, audio and visual equipment engineers, in their pursuit for highly stable motor speeds, have since shifted their attention to brushless d.c. motors used in conjunction with a feedback circuit.

(1) *Torque.* To examine $\oint_\theta \boldsymbol{B} \cdot \mathrm{d}\boldsymbol{H}$ in eqn (8.54), we express it in the following form:

$$\oint_\theta \boldsymbol{B} \cdot \mathrm{d}\boldsymbol{H} = \oint_\theta B_\theta \, \mathrm{d}H_\theta + \oint_\theta B_r \, \mathrm{d}H_r + \oint_\theta B_z \, \mathrm{d}H_z. \qquad (8.60)$$

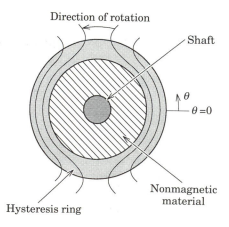

FIG. 8.11. Magnetic flux distribution in a two-pole hysteresis motor.

Only the first term need be considered, for the following reason. In Fig. 8.11 we show the cross-section of a two-pole hysteresis motor. Since the magnetic field consists of θ- and r- but no z-components, the last term of eqn (8.60) becomes zero. Furthermore, if the steel ring where hysteresis is occurring is thin, the θ-component of magnetic flux is the strongest of the three components, making the first term the most important one. Now we assume that the magnetic field pattern in Fig. 8.11 revolves in the counterclockwise (or positive θ) direction, and is given by

$$H_\theta = H_\mathrm{M}\cos(\omega t - \theta + \varphi_0) \tag{8.61}$$

where φ_0 is some appropriate phase angle.

If we assume that the rotor is at a standstill, the above equation represents the fluctuating magnetic field in the rotor. For a set θ, the magnetic field oscillates with a frequency $f = \omega/2\pi$. So at any constant θ value, a hysteresis such as shown in Fig. 8.9a exists. However, the location along the hysteresis loop depends on the angle θ. If we thus plot the (B_θ, H_θ) relationship for a given point in time, a loop such as shown in Fig. 8.12 is obtained. Although we omit the details, it is known that when θ is varied from 0 to 2π, the corresponding point on the loop will travel

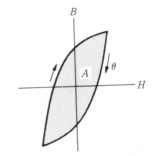

Fig. 8.12.   (B, H) curve with space-coordinate parameter.

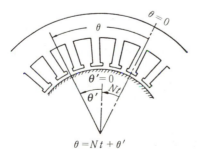

Fig. 8.13.   Relation between stator coordinate θ and rotor coordinate θ'.

clockwise. This is opposite to the counterclockwise movement (for increasing t) along the loop in Fig. 8.9.

Similar results are obtained when the rotor is turning at a speed less than the synchronous speed. From Fig. 8.13, θ and θ' have the following relationship:

$$\theta = Nt + \theta' \tag{8.62}$$

where θ and θ' are chosen to coincide at $t = 0$. If we substitute this into eqn (8.61), we obtain the oscillating magnetic field H_θ in the rotor coordinate system as follows:

$$H_\theta = H_M \cos\{(\omega - N)t - \theta' + \varphi_0\}. \tag{8.63}$$

Therefore, for a given θ', the magnetic field displays the same hysteresis loop, albeit at a different angular frequency $(\omega - N)$. So $\oint_\theta B \cdot dH$ yields the same value as above at any given point in time.

Let A be the area of the loop in Fig. 8.12. When H_θ is given by a simple sine wave, the loops in Figs 8.9 (with respect to time) and 8.12 are known to have equal areas. Therefore the torque is given by

$$T = \int_z \int_r rA \, dr \, dz. \tag{8.64}$$

Integration for z is carried out over the rotor's length, while r is integrated from the inner to the outer diameter. Assuming that A is not affected by r or z, the above equation becomes

$$T = \frac{V}{2\pi} A \tag{8.65}$$

where V is the volume of the ring and A the hysteresis area.

The above discussion applies to the two-pole motor. For four- and six-pole motors, the (B_θ, H_θ) hysteresis loop is circumscribed twice and thrice respectively to complete the integration, and the area must be doubled or tripled correspondingly. Therefore, in general

$$T = \frac{pVA}{2\pi} \tag{8.66}$$

where p is the number of pole pairs (for example, $p = 1$ if there are two poles; p is generally equal to half the number of poles).

(2) *Rotor input.* If the number of poles is equal to $2p$, the magnetic fields for the respective coordinate system are given by the following equations:

for the stationary system,

$$H_\theta = H_M \cos(\omega t - p\theta + \varphi_0) \tag{8.67a}$$

for the coordinate system revolving with the rotor,

$$H_\theta = H_M \cos\{(\omega - pN)t - p\theta' + \varphi_0\}. \qquad (8.67b)$$

The periods are then given by

$$t_0 = 2\pi/\omega \qquad (8.68)$$

$$t_0' = 2\pi/(\omega - pN). \qquad (8.69)$$

Therefore using $\oint_{t'} \mathbf{H} \cdot d\mathbf{B} = A$, the loss within the rotor is

$$\langle P_{LOSS} \rangle = \frac{\omega - pN}{2\pi} \int_V A \, dV = \frac{\omega - pN}{2\pi} VA. \qquad (8.70)$$

The rotor input is then obtained as

$$P_{IN} = NT + P_{LOSS} = \frac{NpVA}{2\pi} + \frac{\omega - pN}{2\pi} VA = \frac{\omega}{2\pi} VA. \qquad (8.71)$$

It follows that $\oint_{t'} \mathbf{H} \cdot d\mathbf{B}$ is equal to A in the stator coordinate system as well.

(3) *When higher harmonics exist.* In actual hysteresis motors, the stator's toothed structure causes fluctuations of the magnetic field near the ring's

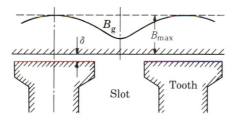

FIG. 8.14. Gap structure and flux fluctuations due to slot-and-tooth structure of stator core.

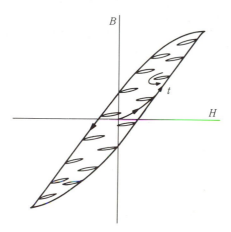

FIG. 8.15. Hysteresis exhibiting minor loops.

surface, as shown in Fig. 8.14. This results in the creation of many minor loops within the hysteresis loop for time (see Fig. 8.15). Since points on the minor loops travel counterclockwise (with increasing time), as in the major loop, the overall rotor loss increases. The hysteresis loop for the space coordinate at the instant the synchronous speed has been reduced, on the other hand, is shown in Fig. 8.16. It shows a reduced area, resulting in a

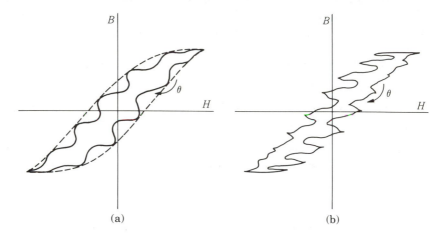

(a)   (b)

FIG. 8.16. *B–H* space-coordinate loop near the synchronous speed: (a) minor loops with zero area; (b) minor loops with finite area.

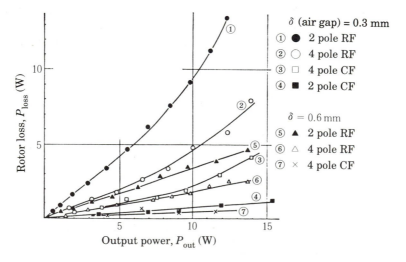

FIG. 8.17. Relation between output at synchronous speed and rotor loss caused by slot harmonics (from ref. 5). Alni cast magnetic material is used for rotor. The coercive force for the circumferential (CF) type is 0.84 A m^{-1} (or 105 Oe), and 1.83 A m^{-1} (or 230 Oe) for the radial flux (RF) type.

lower torque. The larger the minor-loop areas are in the time hysteresis, the smaller becomes the area for the 'space' hysteresis loop.

Figure 8.17 shows experimental results of output and rotor loss for various cases[7]. In the circumferential-flux-type motor, the ring support is made from a non-magnetic material which causes the magnetic field to be distributed primarily in the circumferential direction (see Fig. 8.18a). Most hysteresis motors in use are of this type. The radial flux type shown in Fig. 8.18(b), on the other hand, has a support made from a material with high permeability which distributes the magnetic field in the radial direction.

The following conclusions can be obtained from Fig. 8.17:

(1) rotor loss is smaller for larger gap widths;

(2) circumferential flux motors have lower rotor loss than radial flux types;

(3) in circumferential flux types, a 2-pole configuration generates less loss than a 4-pole one;

(4) In radial flux types, a 4-pole configuration generates less loss than a 2-pole one.

8.5 Comparison of the hysteresis motor and ultrasonic motor

We discussed the hysteresis motor above, deriving equations for torque and loss, so that a comparison could be made with the ultrasonic motor. As we have seen, the hysteresis motor employs the hysteresis effect displayed between B (magnetic flux density) and H (magnetic field strength) in magnetic substances. This relation between B and H closely resembles that between displacement (x) and force (f) in a frictional system as shown

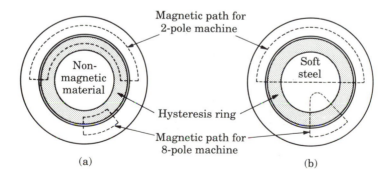

FIG. 8.18. (a) Circumferential and (b) radial flux-type hysteresis motors; radial flux type should have material with high coercive force, since the magnetic path is short.

in Fig. 2.6 (p. 30), in which hysteresis occurs between rotational angle and torque. For this reason, magnetic hysteresis is sometimes called magnetic friction.

Quite possibly, in the ultrasonic motor the microscopic displacements at the rotor-stator interface are subject to complex hysteresis characteristics. Although magnetic and mechanical hysteresis are similar, it does not follow necessarily that the hysteresis motor and ultrasonic motor share common characteristics, and in fact they are quite different in some respects. Even so, the comparison is useful, as it opens up possibilities for applying results of hysteresis motor research, as well as its methodology, to the study of ultrasonic motors.

8.5.1   Expressing ultrasonic motor quantities using the periodic Stieltjes integral

(1)   *Rotor input.* We have shown that the rotor input for an electromagnetic motor, when no current exists in the rotor and torque is created by magnetization only, is given by eqn (8.57). What is the rotor input for the ultrasonic motor? The reader may recall that this was discussed in Chapter 5, where it was given by eqn (5.129). Thus, letting t_0 be the period of the ultrasonic wave, we have

$$P_{IN} = \int_S \frac{1}{t_0} \int_t^{t+t_0} f_x \frac{du}{dt} \, dt \, dS \tag{8.72}$$

where t_0 is the vibrational period of a particle at the stator–rotor contact surface, f_x the x-component of the surface force exerted by the stator on the rotor (see Fig. 8.19), and u the displacement in the x-direction of a particle on the stator surface from its initial position.

Considering that

$$\frac{du}{dt} \, dt = du \tag{8.73}$$

we can write P_{IN}, using the periodic Stieltjes integral with parameter t, as

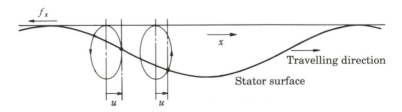

FIG. 8.19.   Contact surface of the travelling-wave ultrasonic motor.

$$P_{\text{IN}} = \int_s \frac{1}{t_0} \oint_t f_x \, du \, dS. \qquad (8.74)$$

This corresponds to eqn (8.57). Whereas eqn (8.57) is a volume integral, eqn (8.74) is a surface integral. This is an important difference.

(2) *Loss*. The loss within the rotor was given by eqn (8.58), then expressed in the form of eqn (8.59). Assuming that the rotor is a rigid body, the loss for the ultrasonic motor can be expressed in a similar form. When a particle on the contact surface remains stationary with respect to the rotor, there is no frictional loss. If on the other hand the contact surface has a finite du/dt value, the loss is obtained by multiplying this value by the frictional force f. Thus

$$P_{\text{LOSS}} = \int_s \frac{1}{t_0'} \int_t^{t+t_0'} f_x \left(\frac{du}{dt}\right)' dt \, dS$$

$$= \int_s \frac{1}{t_0'} \oint_t{}' f_x \, du \, dS \qquad (8.75)$$

where t_0' is the vibrational period in a coordinate system in motion with the rotor; ( )′ and ∫′ are the derivative and integral, respectively, with respect to time in this coordinate system.

8.5.2 Expressions for torque and thrust

The mechanical output is obtained by subtracting P_{LOSS} from P_{IN}. Refer to Fig. 8.20(a). In a linear motor, this is equal to the product of thrust F and velocity v. So

$$P_{\text{OUT}} = Fv = P_{\text{IN}} - P_{\text{LOSS}} = \int_s \frac{1}{t_0} \int_t^{t+t_0} f_x \left\{\left(\frac{du}{dt}\right) - \left(\frac{du}{dt}\right)'\right\} dt \, dS.$$

$$(8.76)$$

Since $(du/dt)'$ represents the speed of a particle on the stator surface as seen from a coordinate system with a constant velocity v,

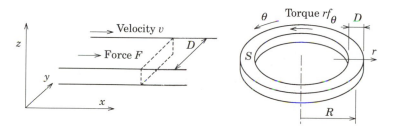

FIG. 8.20.   Thrust and torque.

$$\left(\frac{du}{dt}\right)' = \left(\frac{du}{dt}\right) - v. \tag{8.77}$$

Therefore,

$$Fv = \int_S \frac{1}{t_0} \int_t^{t+t_0} f_x v \, dt \, dS = v \int_S \frac{1}{t_0} \int_t^{t+t_0} f_x \, dt \, dS. \tag{8.78}$$

Dividing this equation by v and letting t_0 approach zero, assuming that f_x is a continuous function in time, we obtain the instantaneous thrust F:

$$F = \int_S f_x \, dS. \tag{8.79}$$

For a rotary motor, the torque, using the radial coordinate r as shown in Fig. 8.20(b), is given by

$$T = \int_S r f_\theta \, dS. \tag{8.80}$$

   Equation (8.79) states the obvious fact that the thrust is obtained by integrating the tangential forces over the surface, and serves as a check on the expressions derived. (Or in the case of eqn (8.80), that torque is obtained by taking the surface integral of the product of the tangential force f_θ and r.) Torque for the hysteresis motor was given by eqn (8.54). We can derive a similar expression for the ultrasonic motor. If, for example, P_{LOSS} is zero, eqn (8.76) becomes

$$P_{\text{OUT}} = Fv = \int_S \frac{1}{t_0} \int_t^{t+t_0} f_x \frac{du}{dt} \, dt \, dS. \tag{8.81}$$

Since

$$\frac{du}{dt} = \frac{du}{dx}\frac{dx}{dt} \tag{8.82}$$

and dx/dt is the phase velocity v_{ph} of the travelling wave (see Fig. 8.21),

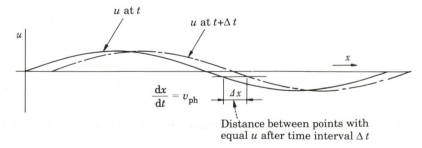

FIG. 8.21.   Illustration of why dx/dt is equal to the phase velocity v_{ph}.

$$Fv = v_{\text{ph}} \int_S \frac{1}{t_0} \int_t^{t+t_0} f_x \frac{du}{dx} \, dt \, dS \qquad (8.83)$$

$$Fv = v_{\text{ph}} \int_S f_x \frac{du}{dx} \, dS = v_{\text{ph}} \int_y \int_x f_x \frac{du}{dx} \, dx \, dy$$

$$= v_{\text{ph}} \int_y \left(\oint_x f \, du \right) dy. \qquad (8.84)$$

If this function is constant along the rotor's width (i.e. in the y-direction), then

$$Fv = v_{\text{ph}} D \oint_x f \, du. \qquad (8.85)$$

We can express v_{ph} in terms of wavelength λ and frequency v as

$$v_{\text{ph}} = \lambda v. \qquad (8.86)$$

For a rotary motor, the wavelength λ is equal to $2\pi R/p$, so

$$T = \frac{2\pi v R D}{pN} \oint_\theta f \, du \qquad (8.87)$$

where N is the rotor's speed.

8.5.3 Elastic-body rotor

We next examine the case where the rotor consists of an elastic material and deformation in the tangential direction must be considered. If we let u_R be the tangential (or shearing) displacement of a particle at the rotor's surface, then P_{LOSS} is

$$P_{\text{LOSS}} = \int_S \frac{1}{t_0'} \int_t^{t+t_0'} f_x \left\{ \left(\frac{du}{dt} \right)' + \left(\frac{du_R}{dt} \right)' \right\} dt \, dS. \qquad (8.88)$$

Using eqns (8.72) and (8.77), we obtain

$$P_{\text{OUT}} = vF = P_{\text{IN}} - P_{\text{LOSS}} = v \int_S f \, dS - \int_S \frac{1}{t_0'} \oint_t' f \, d \, u_R \, dS \qquad (8.89)$$

so

$$F = \int_S f \, dS - \frac{1}{v} \int_S \frac{1}{t_0'} \oint_t' f \, du_R \, dS \qquad (8.90)$$

where t_0' is the period of the rotor's vibrational deformation as seen from the rotor.

Next, if we consider the case when pressure causes concave deformations

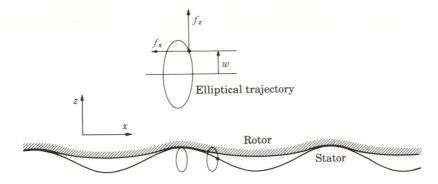

FIG. 8.22. Rotor deforms in the z-direction also.

at the rotor's contact surfaces while it rotates, as shown in Fig. 8.22, we obtain the following equations:

$$P_{\text{IN}} = \int_S \frac{1}{t_0} \int_t^{t+t_0} \left\{ f_x \left(\frac{du}{dt} \right) + f_z \left(\frac{dw}{dt} \right) \right\} dt \, dS \tag{8.91}$$

$$P_{\text{LOSS}} = \int_S \frac{1}{t_0'} \int_t^{t+t_0'} \left[f_x \left\{ \left(\frac{du}{dt} \right)' + \left(\frac{du_R}{dt} \right)' \right\} + f_z \left(\frac{dw}{dt} \right)' \right] dt \, dS \tag{8.92}$$

where w is the displacement of a particle in the z-direction and f_z the force in the z-direction (see Fig. 8.22).

The thrust is given by

$$F = \frac{P_{\text{OUT}}}{v} = \frac{P_{\text{IN}} - P_{\text{LOSS}}}{v} \tag{8.93}$$

$$F = \int_S f_x \, dS - \frac{1}{v} \int_S \frac{1}{t_0'} \oint_t' f_x \, du_R \, dS - \frac{1}{v} \int_S \left(\frac{1}{t_0'} \oint_t' f_z \, dw - \frac{1}{t_0} \oint_t f_z \, dw \right) dS. \tag{8.94}$$

We shall examine the last term in this equation. We assume that $\oint f_z \, dw$ and $\oint_t' f_z \, dw$ are equal. Now, t_0' is the vibrational period of a particle at the rotor's contact surface with respect to the rotor's moving coordinate system, while t_0 is the period of a particle at the stator's contact surface from the perspective of the stator coordinate system. Since the rotor and the travelling wave move in opposite directions, the frequency at the rotor's surface (v') is higher than that for the stator's surface (v). Thus v' is given by

$$v' = \frac{1}{t_0'} = \frac{1}{t_0} + \frac{v}{\lambda} \tag{8.95}$$

where

$$\frac{1}{t_0} = v. \tag{8.96}$$

Therefore

$$\frac{1}{t_0'} - \frac{1}{t_0} = \frac{v}{\lambda}. \tag{8.97}$$

Substituting this into eqn (8.94), we obtain for the thrust

$$
\begin{aligned}
F &= \int_s f_x \, dS - \frac{1}{v} \int_s \frac{1}{t_0'} \oint_t' f_x \, du_R \, dS - \frac{1}{\lambda} \int_s \oint_t f_z \, dw \, dS \\
&= \int_s \left\{ f_x - \frac{1}{v} \left(v + \frac{v}{\lambda} \right) \oint_t' f_x \, du_R - \frac{1}{\lambda} \oint_t f_z \, dw \right\} dS
\end{aligned}
\tag{8.98}
$$

or

$$
= \int_s \left\{ f_x - \frac{v}{v} \oint_t' f_x \, du_R - \frac{1}{\lambda} \left(\oint_t' f_x \, du_R + \oint_t' f_z \, dw \right) \right\} dS. \tag{8.99}
$$

8.5.4 The need for further investigations

The equations above show that the thrust (or torque) is lower than the force exerted on the rotor by the stator. This can be accounted for by elastic deformations at the rotor's surface. As we mentioned in the last chapter, the rotor, rather than being a rigid body, requires a certain amount of elasticity in practice. Thus the rotor's elasticity is a 'necessary evil', which must be handled carefully in motor design. The stator's comb-tooth structure is very likely an important factor in this respect. Earlier in this chapter we discussed the effects of the stator's tooth-structure in the hysteresis motor. A similar investigation for the ultrasonic motor is necessary.

8.6 A comparison with d.c. motors

We shall now change our approach and compare the ultrasonic motor with the d.c. motor. Such a comparison will not only help to point out the inherent characteristics of the ultrasonic motor but also suggest possible areas for future research. The basic construction of the d.c. motor is shown in Fig. 8.23. Brushes and commutators, shown in the figure, play important roles in the d.c. motor, although details of their workings are outside the scope of this text. The d.c. motor functions much like the brushless d.c. motor, which in turn is similar in construction to the synchronous motor (a type of a.c. motor), possessing a permanent magnet as the rotating body

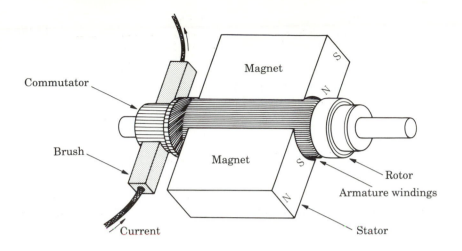

Fɪɢ. **8.23.**   Basic construction of d.c. motor.

in Fig. 8.3. Thus the basic theory we have discussed so far applies also to the d.c. motor.

The d.c. and brushless d.c. motors differ in one respect, however: whereas the latter uses an electronic circuit with transistors etc. to create a rotating magnetic field, the former uses a mechanical switching mechanism with brushes and commutators. When an electronic circuit is used, the winding remains stationary while the permanent-magnet rotor turns; when brushes and commutators are used, the permanent magnet is fixed while the winding rotates. For a detailed discussion on these subjects, see ref. 8.

8.6.1   Current and voltage in the equivalent circuit

We shall compare the equivalent circuits of the two types of motor. A d.c. motor with a simple load and loss can be represented by the equivalent circuit in Fig. 8.24[9]. In this circuit, current and voltage have the following meanings.

Current. Torque is proportional to the current i, where the proportionality constant is called the torque constant K_T:

$$T = K_T i. \tag{8.100}$$

The input current i is further split up into i_D, i_R and i_L, where:

$K_T i_D$ is the torque required to overcome the shaft's friction, wind resistance, and retarding torque associated with iron loss;

$K_T i_R$ is the torque required to accelerate or decelerate the rotor;

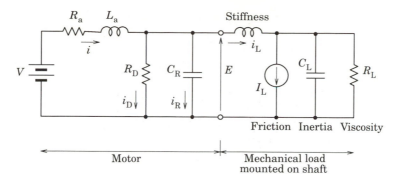

FIG. 8.24. Equivalent circuit of d.c. motor and load: R_a, winding resistance; L_a, winding inductance; R_D, mechanical loss and iron loss resistance; C_R, rotor inertia.

$K_T i_L$ is the torque for driving the load.

Voltage. The terminal voltage at R_D or C_R is proportional to the rotational speed N. Using K_E (the back electromotive force constant) as the proportionality constant, we have

$$E = K_E N. \tag{8.101}$$

In SI units or any other self-consistent unit system, K_E and K_T are equal.

The circuit elements represent the following mechanical quantities:

Capacitance (C): moment of inertia (or inertia)

Reactance (L): spring effect (or stiffness) of the shaft and load

Resistance (R): viscosity

Current sink (I): friction

We shall note a difference that is seen when compared with the equivalent circuit discussed in Chapter 2, which presents mechanical relations between force and velocity, and is suited to deal with the ultrasonic motor as seen in Chapter 3. In such a circuit the source is force or torque, and it is proportional to the consumed current when a d.c. motor is used as the torque source. A mass or moment of inertia was represented by an inductance, and a spring by a capacitance. In the equivalent circuit having a voltage source as discussed here, inertia is represented by a capacitance, and the spring's stiffness by an inductance. Another difference occurs in the way of presenting frictional load. In a mechanical circuit in Chapter 2 it is presented by a normal or Zener diode, while a current sink is suitable for representing friction in a voltage source circuit. It may be understood in such a way that the motor will not move unless the torque available to the load (which is presented by $K_T i_L$) exceeds the static friction, which is K_T multiplied by the sink current. Refer to ref. 8 for detailed discussions on equivalent capacitance and inductance in this sort of circuit.

For the ultrasonic motor, the equivalent-circuit parameters and variables represent different quantities. In Chapter 3, we showed that

(1) current is proportional to the motor's speed;

(2) voltage is proportional to the motor's torque.

Thus current and voltage in equivalent circuits represent opposite dynamic quantities in the two motors. The quantities represented by elements C and L are also reversed for the ultrasonic motor:

(1) capacitance represents stiffness (a property of the spring);

(2) inductance represents moment of inertia or mass.

The constants K_T and K_E for the d.c. motor correspond to the force factor A. In the ultrasonic motor, the voltage applied to the piezoelectric ceramic multiplied by the force factor A gives the force F_i generated within the ceramic element. If we subtract from this the loss due to internal impedance, we obtain the force available externally (to drive the load) (see eqn (3.10), p. 55). The ceramic's vibrational velocity v multiplied by A, on the other hand, gives the current I_s required to cause deformation in the ceramic element. Or,

$$F_i = A V_i \tag{8.102}$$

$$I_s = A v. \tag{8.103}$$

This is shown again in the equivalent circuit of Fig. 8.25. Since the blocking capacitance C_d can be cancelled by placing an inductance L_d in parallel, we can substitute the simple equivalent circuit of Fig. 8.26(b). In this circuit,

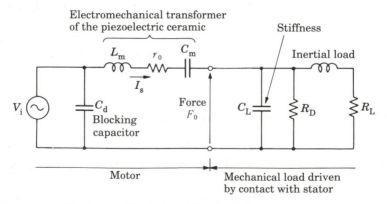

FIG. 8.25.   Equivalent circuit of ultrasonic motor.

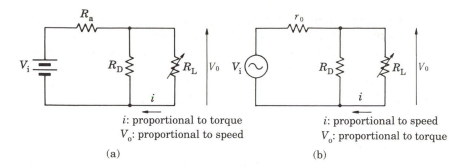

i: proportional to torque
V_0: proportional to speed

i: proportional to speed
V_0: proportional to torque

(a)

(b)

FIG. 8.26. Comparison of equivalent circuits simplified for steady-state conditions: (a) d.c. motor; (b) ultrasonic motor.

the equivalent resistance R_D represents loss at the contact surface (discussed in Chapter 6).

8.6.2   Torque–speed characteristics and conditions for maximum efficiency

We shall now examine the relation between torque, speed and efficiency for the d.c. and ultrasonic motors, using the equivalent circuits. As we are not concerned with dynamic characteristics at the moment, we can omit L_a and let C_R be infinite in Fig. 8.24 to obtain Fig. 8.26(a). For the ultrasonic motor, we can use circuit (b) by assuming that the motor is driven at the resonant frequency so that C_m and L_m will cancel each other. Thus we have two equivalent circuits (a) and (b), which are similar in form, but in which current and voltage represent opposite quantities, as we saw above.

Figures 8.27(a) and (b) show graphs of current against voltage for the two equivalent circuits. Note that the conditions for obtaining maximum efficiency are different for the two cases. The smaller R_a or r_0 is in comparison with R_L (which represents output), the higher is the efficiency. Also, the higher is R_D (representing iron loss for the d.c. motor, or sliding loss for the ultrasonic motor), the more efficient is the motor. Thus efficiency improves when

$$R_a \, (\text{or } r_0) \ll R_D. \tag{8.104}$$

To achieve maximum efficiency under this condition, R_L is given by

$$R_L \simeq \sqrt{R_a R_D} \tag{8.105}$$

while the maximum efficiency $\eta_{\max}$ and motor speed N^* are given by

$$\eta_{\max} = \frac{M-1}{M+1} \tag{8.106}$$

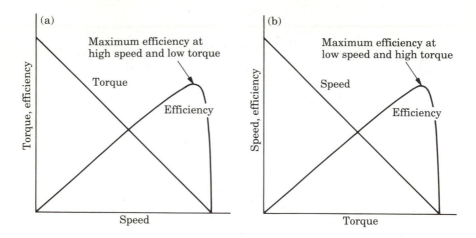

Fig. 8.27. Comparison of torque vs. speed and efficiency characteristics for ideal motor: (a) d.c. motor; (b) ultrasonic motor.

$$N^* = \frac{M}{M + 1} N_0$$ (8.107)

where

$$M = \sqrt{[(R_a + R_D)/R_a]}$$ (8.108)

and N_0 is the no-load speed.

From these equations, we reach the following conclusions:

(1) for the d.c. motor, the maximum efficiency occurs near the no-load speed (i.e. at low torque);

(2) for the ultrasonic motor, this occurs at low speed and high torque.

Thus the d.c. motor is suited for high-speed operation, while the ultrasonic motor is effective at low speed.

8.6.3   Consideration of dynamic characteristics

We have discussed how high efficiencies are achieved under steady-state operation. However, certain limitations of the d.c. motor become apparent when dynamic characteristics are considered, which in turn suggests possible applications for the ultrasonic motor. As an example, we shall consider the d.c. motor used for power steering in automobiles. Although this is a complex mechanism which requires more than a few pages to be fully explained, we focus here on the motor's inertia.

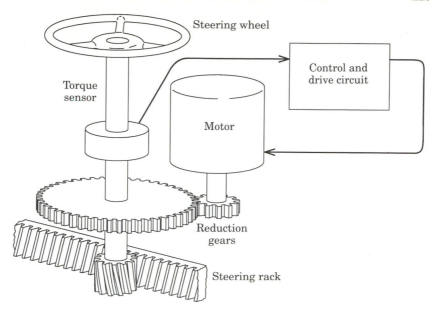

FIG. 8.28. Basic components of power-steering mechanism.

The model shown earlier in Fig. 2.28 illustrates the principle of a power-steering mechanism. Figure 8.28 shows a system in which the motor generates an auxiliary torque in response to the driver's manually applied torque to the steering wheel; Fig. 8.29 is the equivalent circuit. Since the d.c. motor is most efficient when operated-around the no-load speed, a

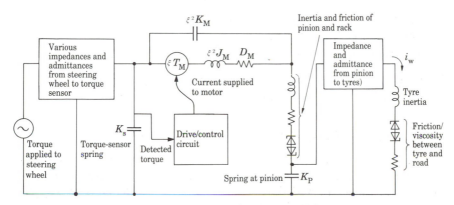

FIG. 8.29. Mechanical equivalent circuit for power-steering mechanism; time integral of i_w is the wheel's steering angle. T_M, motor torque; J_M, Motor inertia; K_M, Spring constant at reduction gears; D_M, Equivalent resistance representing viscosity at reduction gears; ξ, Inverse of reduction ratio.

reduction gear is used to create the low speed and high torque necessary. This also greatly increases the motor's effective moment of inertia to the steering wheel, which is equal to the motor's moment of inertia times the square of the reduction ratio (i.e. of the order of 10). Whenever such large inertias are involved in the system, not only does the control response worsen but a low-frequency resonance is likely to occur owing to the compliance with interlocking peripheral parts. Therefore a mechanism to disconnect the motor's mechanical linkage when not in use — a clutch or moving coil for example — may be necessary.

This limitation is an inherent characteristic of the d.c. motor, as well as of other electromagnetic motors, and cannot be removed by improvements in the motor's structure. So the ultrasonic motor has advantages when considered from the viewpoint of dynamic characteristics. Its inherent structure allows it to achieve low speed coupled with high torque without a speed reduction gear, and its equivalent inertia is small.

8.6.4   *Drawbacks of the ultrasonic motor*

The ultrasonic motor, on the other hand, also has shortcomings which must be solved successfully before practical applications can be realized. The difficulties lie at the frictional contact surface between stator and rotor. There are two problems that need to be solved.

The first problem is actually the other side of the coin of the advantage we cited in the example above of the steering wheel mechanism. For easy recovery of the normal steering position it is desirable that the motor should be able to rotate freely under the external torque when electric power is not supplied. This is not the case for the ultrasonic motor. Although a clutch mechanism at the rotor–stator contact surface would solve this problem, it would make the structure too complicated.

The switched reluctance motor[10], another type of electromagnetic motor, has a structure well suited in this respect. As shown in Fig. 8.30, a switched reluctance drive is basically a VR-type stepping motor that is used as a brushless d.c. motor. Its holding torque is zero when no power is supplied, and thus it is able to rotate freely. The permanent-magnet brushless d.c. motor can also rotate freely when no power is supplied, as long as current is prevented in the stator circuit. If by error the stator circuit is shorted, however, the permanent magnet will create a current and exert a braking torque on the rotor.

The other problem is loss created at the contact surface. If the two contact surfaces mesh like gears, as in Fig. 5.26, no sliding occurs. Then, at low speeds, the stator's vibrational amplitude is small and the equivalent electrical impedance of the load is large, resulting in a low consumption of current. In this case, power consumption is small and so the efficiency is high. However, as we have seen in Chapters 6 and 7, the low-speed efficiencies achieved in actual ultrasonic motors are not at a satisfactory

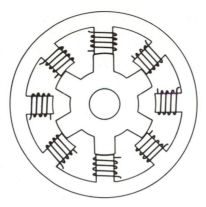

Fɪɢ. 8.30. Basic structure of the switched reluctance motor; rotor and stator both employ soft-iron cores.

level. This is because, as we examined in Chapter 5, both the frictional loss which accompanies sliding at the contact surface and the loss due to internal viscosity of the elastic body (or rotor) increase at low speeds.

8.7 Concluding remarks

In this volume, we have discussed the ultrasonic motor, from a basic introduction to practical design aspects with an emphasis on theory. We have seen that low speed and high torque characterize the ultrasonic motor when compared with common electromagnetic motors. In order to fully utilize this feature, in addition to extending the motor's life-span and improving its reliability, basic research of the behaviour at the stator–rotor contact surface is necessary. It is known that under a high vacuum, metal contact surfaces display lubrication and friction properties quite different from those displayed under normal atmospheric pressure or partial vacuum. This shows that contact behaviour is related to environmental (atmospheric) conditions and thus suggests the need for further research in materials science.

The authors believe that the ultrasonic motor's theory and technology have been given a fairly extensive coverage in this text from a macroscopic viewpoint. For further development, it seems that microscopic considerations will become essential.

References

1. Kenjo, T. (1991). *Electric motors and their controls—an introduction*. Oxford University Press.

2. Livens, G. H. (1918) *The theory of electricity*. Cambridge University Press.
3. Poynting, J. H. (1884). On the transfer of energy in the electromagnetic field. *Philosophical Transactions of the Royal Society of London*, **175**, Part II.
4. Kenjo, T. (1975). Rotating AC machinery theory based on Poynting's theorem. *Journal of the Institute of Electrical Engineers of Japan*, **95**, 359–66.
5. Teare, B. R., Jr (1937). Theory of hysteresis-motor torque. Unpublished thesis. Yale University.
6. Teare, B. R., Jr (1940). Theory of hysteresis-motor torque. *AIEE Transactions*, **59**, 907–12.
7. Kenjo, T. (1970). Slot harmonics loss of hysteresis motor. *Journal of the Institute of Electrical Engineers of Japan*, **90**, 2344–53.
8. Kenjo, T. and Nagamori, S. (1985). *Permanent-magnet and brushless DC motors*. Chapters 1–4, Clarendon Press, Oxford.
9. ibid. Chapters 6 and 7.
10. Miller, T. J. E. (1989). *Brushless permanent-magnet and reluctance motor drives*. Clarendon Press, Oxford.

Autobiographical notes

(Toshiiku Sashida)

I was born in Tokyo in January 1939, three years before Japan entered World War II. When the Allies started the air raids on Tokyo, my family had already been evacuated to the countryside. When the war ended in 1945, I was in the first grade of elementary school. By that time, most of Japan's major cities had been decimated by the B-29's incendiary bombs. Tokyo had been hit particularly hard and only a few scattered buildings, including the Diet Building, remained standing.

At the local town close to where we had been evacuated, there was an airfield. Almost all the planes had been destroyed during the war, except for a few that had been hidden in a nearby wood. These planes did not remain for long either, since the local townspeople salvaged whatever they could, following the collapse of military rule. Along with my older brothers, I too went to the wood to look for parts that could be used as toys. I found what was probably a panel gauge and, after taking it home, took it apart to examine the inside.

My family returned to Tokyo a year after the war ended. There, my father started a small retail food business. Although I liked making things with my hands, there were no ready-made kits available at the time. If a boy wanted to make something, he had to start from scratch using a penknife or a pair of tinsnips. By the time I was in fourth grade, however, a few model shops had opened up in town. I often visited one of these shops just to look at the many items on display, and it wasn't long before a small motor caught my fancy. I was not in the habit of begging things from my father, but I couln't give up this motor and eventually he granted my wish. It was a d.c. motor with a U-shaped magnet.

Although I had ambitious plans to make a mechanical doll with this motor, that never materialized; I kept taking the motor apart and putting it back so many times that eventually some parts were lost. The salvaged panel gauge and this motor still remain as pleasant childhood memories for me.

After junior high school, I intended to go to a regular senior high school. So I was surprised to discover on my first day of classes that I had actually enrolled in a technical high school. In fact, I hadn't even been aware that such schools existed and had assumed that regular high schools taught electrical circuitry and machine design. So it was by a lucky chance that I found myself in a school which taught what I was most interested in. Here, I studied telecommunication science and learned about a.c. circuits and electronic circuits for vacuum tubes. This knowledge came in handy when years later I designed the circuitry for the ultrasonic motor.

After graduating from senior high school, I enrolled at the Science University of Tokyo, where I majored in physics. In lectures on theoretical physics, I was introduced to quantum mechanics. Despite my efforts to master this subject, I could not come to terms with its abstractness. I could see Newtonian forces at work in everyday phenomena, but I felt uncomfortable with quantum mechanics' need to use imaginary numbers in describing the real world. Thus I found myself leaning to the experimental side of physics.

For my graduation thesis, I chose a topic in semiconductor research: in technical terms, on the effects of crystal lattice defects on electric conductivity. I used to stay at my laboratory to conduct experiments late into the night, my thoughts absorbed in the mystery of the electron's behaviour in semiconductors. I soon found myself 'thinking like an electron'. I felt that, through experiments, I had somehow succeeded in taking a brief glimpse into the realm of the atom, governed by laws of quantum mechanics.

As graduation approached, I thought that I should work for a large company, like any other self-respecting college graduate. In those days, a senior spent his summer working as a trainee for a prospective employer and this substituted as a hiring test. During my senior year, I found summer 'work' at a well-known telecommunications and computer firm. Although I felt that I had learned quite a great deal during this period, I was disappointed next autumn when notified that I had not been hired. My co-author, Professor Kenjo, thinks that Japanese firms at the time were still too traditionally minded to accept creative personalities on to their payrolls.

It was not the end of the world, however, and I stayed busy with my experiments. My thesis adviser, who saw me absorbed in experiments, offered me a position as his assistant, which I felt I had to decline because of the extremely sorry condition of the laboratory building. However, with his introduction, I did find a job as an assistant in the strength-of-materials laboratory in Yokohama National University's Mechanical Engineering Department. It was during my three years there that I learned about ultrasonic vibrations and applied mechanics, which laid the foundation for later work on the ultrasonic motor. It was here that I also learned academic research methodology.

By the end of three years, I was starting to find the university research environment confining. As an assistant, I wasn't allowed to expand research outside the established areas set by the laboratory's director. It was about that time that I was introduced to Mr Kobayashi, the owner of a small factory.

During the two and a half years that I worked at Mr Kobayashi's factory, I gained some invaluable experience. There weren't ICs or microprocessors yet, and my duties consisted in designing and assembling transistor circuits for machine-tool automation. So I learned the practical aspects of what is now known as factory automation or numerically controlled machines. Mr Kobayashi also taught me the business end of running a small firm. This was very different from the content of a university management course, nor had it anything to do with modern business strategies for mergers and sellouts. Rather, it was the art of persevering through hard times, much like an animal in hibernation waiting for the arrival of spring. Although I was hired with our mutual understanding that eventually I would branch out on my own, it was still very sad for both of us when the time came for me to leave.

In 1968 I started up a firm and named it Shinsei Industries. 'Shinsei' means 'renewal' in Japanese, and the name signified a fresh start. Finding a customer is all-important, especially when one is just starting a business. The first order we got, through a relative, was to construct a hot-press machine for welding conveyor belts. Somewhat later, my former employer Mr Kobayashi introduced us to a customer who wanted us to manufacture lapping machines. These are machines used to grind an object's surface to produce a flat and smooth finish. Working with lapping machines, I learned how machines with high precision are created from ones of lower precision. This process of creating machines with ever-higher precision is fundamental to technological advancement. The lapping machine is an example of this principle and, although I was not aware of it at the time, the knowhow gained through this work proved invaluable to the development of the ultrasonic motor.

As work at our factory settled into a routine, I came to realize a deficiency in the motors being used in machine tools: there were no motors that were small and at the same time powerful.

In 1974 we received a request to manufacture an 'auto-handler', a robot hand. I accepted the order as a matter of routine. However, from the very start we ran into one difficulty after another. Although our objective was to construct a robot capable of simulating the human hand's movements, our efforts produced nothing that approached that effect. During this period I often looked at my hand and couldn't help but feel wonder at the complexity God had built into it.

Ever since this order, I kept perusing medical books for information on the human hand in my desire to create a human-like robot hand. It became apparent to me that the power source or actuator to substitute for hand muscles was the critical problem. A motor is a type of actuator. Robot actuators need to be compact, light in weight, and powerful, and yet there were no actuators that were driven by traditional means and could compare with the human hand in terms of effectiveness. I realized then that I had found a worthwhile challenge as a scientist and an engineer.

At the time, in addition to the lapping machines, we were developing a special type of machine which used ultrasonic waves. When I did some research for this machine, I chanced on a paper by Professor Kiju Kikuchi of Tohoku University. The paper, entitled 'Theory of the ultimate output of the magnetostriction vibrator', concluded that a generator using magnetically induced ultrasonic vibrations was capable of an energy output of 475 watts per square centimetre. This meant that an energy equivalent to the heat output of a 500 watt electric space heater was generated from an area equal to two pencil cross-sections. Only, in the case of the vibrator, the output took the form of ultrasonic vibrations instead of heat. This was an incredibly high value, impossible to duplicate using electric motors. I resolved then to develop a new type of motor that would utilize ultrasonic waves.

After reading Professor Kikuchi's paper, I started looking for ways to convert ultrasonic vibrations into unidirectional motion usable in an actuator. Although a few mechanisms for converting reciprocating motion into one-way motion were known at the time, they were employed for vibrations of several tens of hertz or less, and amplitudes of at least 1 millimetre. For ultrasonic vibrations, which have frequencies above 20 000 Hz and amplitudes less than a hundredth of a millimetre, they were useless.

After some time I hit upon a principle for producing unidirectional motion, with clues from a Russian scientific paper. My idea was to take advantage of mechanical resonance to improve the efficiency of the electromechanical energy conversion. After much hard work and several prototypes, I developed a small powerful motor, shown in Fig. 4.5(a). A microscope has been attached to check whether the microscopic vibrations have indeed been converted into unidirectional movement.

I was fortunate to have a workshop equipped with machine tools as well as having the skills necessary to build whatever devices or parts that were needed for my research. I probably would not have been able to conduct research in the same way at a university. In research, one must be able to channel all of one's energy into the subject in hand and, whenever an idea takes form, one must be able to implement it immediately to put it to test. The delicate adjustments necessary when machining parts must be done by hand. I had to build and discard hundreds of parts before I was able to make a successful model. If it had been necessary to contract out the parts to some machine shop, not only would it have caused years of delay, but I might have been forced to abandon the project entirely.

After the wedge-type motor, I developed the idea of the travelling-wave motor from clues obtained from Lord Rayleigh's research. Since this is discussed in some detail in Chapter 1, I do not repeat it here.

During the ultrasonic motor's development, I would not have blamed my employees if they had thought of me as some eccentric inventor who was using his business to support an expensive hobby. Even now, I avoid lecture engagements as much as possible and devote most of my time working to improve my invention. When Professor Kenjo invited me to deliver a special lecture for Motortech Japan in 1983, however, I did accept. The lecture created quite a stir at the time, and when we started to receive serious enquiries from many firms, I noticed that my employees started to see me in a somewhat different light.

I should mention how my collaboration with professor Kenjo in writing this book came about. Ever since the early 1980s, when the travelling-wave motor concept was taking shape in my mind, I had been looking for a way to re-examine and organize a theory for electric machines that utilized microscopic vibrations. However, it was almost a decade later, when I had finally succeeded in producing a working model for a robot in a nuclear power plant, that Professor Kenjo gave me the opportunity I had been looking for by contacting me and asking whether I was interested in writing a book. I accepted his offer immediately, since it also gave me a chance to summarize my work up to that point. I was always impressed by his capacity for clear thinking whenever, during our collaboration, we discussed the subject matter. I was also very much impressed by his organizational skills which made it possible to complete the book in a surprisingly short time.

Mr H. Takahashi, who belongs to Professor Kenjo's laboratory, was very helpful when we were designing the manufacturing process of the travelling-wave motor, to be used as the motor for the head support for a Toyota passenger car. He was responsible for designing the hardware and software of the computer control system for automatic testing.

In the Japanese language, the word 'tsukuru' can be used to mean 'to create', 'to make a prototype', or 'to manufacture', while different Kanji (or Chinese characters) are used to distinguish the subtle variations in meaning. Professor Kenjo tells me that he knows of no other instance of a single person who created, prototyped, and mass-produced a new type of motor. While this may be true, I believe that it is not unconnected with my living in Japan during a period of great changes, including defeat at war, the hardships that followed, then economic growth and a measure of prosperity.

In my view, an inventor should not remain satisfied with merely having invented something. I feel that he/she has the social obligation to develop his/her original invention into a product that will actually benefit people. Thus my work, which has gone through several stages of prototype development to make a final product adaptable to mass-production technologies, is simply my personal philosophy put into action.

In closing, I would like further to add my ideas about innovation. As is often said, since Japan started industrializing at a late state, she has built up her industries by importing technology invented in the West and improving upon them. I admit that until quite recently neither the Government nor private entrepreneurs have placed much emphasis on fundamental research. This trend may have had something to do with a few recent cases in which Japanese firms have been accused in court of patent-right infringements. As to whether they were in fact guilty or not, I personally do not agree with such business practices. At the same time, however, I feel that perhaps the inventor should have developed his/her invention to a truly practical level.

Index

Note: Figures and Tables are indicated by *italic page numbers*